IEEE 802.11™ Handbook

...A Designer's Companion
Second Edition

Bob O'Hara
Al Petrick

Published by
Standards Information Network
IEEE Press

Trademarks and Disclaimers

IEEE believes the information in this publication is accurate as of its publication date; such information is subject to change without notice. IEEE is not responsible for any inadvertent errors.

Library of Congress Cataloging-in-Publication Data

O'Hara, Bob, 1956-

The IEEE 802.11 handbook : a designer's companion / authored by Bob O'Hara and Al Petrick.-- 2nd ed.

  p. cm.

Includes index.

ISBN 0-7381-4449-5

1. Wireless LANs. 2. IEEE 802.11 (Standard) I. Petrick, Al, 1957- II. Title.

TK5105.78.O43 2004

621.382'1--dc22

2004059605

IEEE
3 Park Avenue, New York, NY 10016-5997, USA

IEEE and 802 are registered trademarks of the IEEE (www.ieee.org/).

IEEE Standards designations are trademarks of the IEEE (www.ieee.org/).

Wi-Fi is a registered trademark of the Wi-Fi Alliance (http://www.wi-fi.org/).

Jennifer McClain Longman, Managing Editor
Linda Sibilia, Cover Designer

Review Policy

IEEE Press/Standards Information Network publications are not consensus documents. Information contained in this and other works has been obtained from sources believed to be reliable, and reviewed by credible members of IEEE Technical Societies, Standards Committees, and/or Working Groups, and/or relevant technical organizations. Neither the IEEE nor its authors guarantee the accuracy or completeness of any information published herein, and neither the IEEE nor its authors shall be responsible for any errors, omissions, or damages arising out of the use of this information.

Likewise, while the author and publisher believe that the information and guidance given in this work serve as an enhancement to users, all parties must rely upon their own skill and judgement when making use of it. Neither the author nor the publisher assumes any liability to anyone for any loss or damage caused by any error or omission in the work, whether such error or omission is the result of negligence or any other cause. Any and all such liability is disclaimed.

This work is published with the understanding that the IEEE and its authors are supplying information through this publication, not attempting to render engineering or other professional services. If such services are required, the assistance of an appropriate professional should be sought. The IEEE is not responsible for the statements and opinions advanced in the publication.

The information contained in IEEE Press/Standards Information Network publications is reviewed and evaluated by peer reviewers of relevant IEEE Technical Societies, Standards Committees and/or Working Groups, and/or relevant technical organizations. The authors addressed all of the reviewers' comments to the satisfaction of both the IEEE Standards Information Network and those who served as peer reviewers for this document.

The quality of the presentation of information contained in this publication reflects not only the obvious efforts of the authors, but also the work of these peer reviewers. The IEEE Press acknowledges with appreciation their dedication and contribution of time and effort on behalf of the IEEE.

To order IEEE Press Publications, call 1-800-678-IEEE.

Print: ISBN 0-7381-4449-5   SP1136

See other IEEE standards and standards-related product listings at:
http://standards.ieee.org/

Dedication

To my wife, Lisa, for all the support she gave while I was writing this book.

-Bob O'Hara

I dedicate this work to my wife and best friend, Patricia, for her love and support throughout the writing of this book.

- Al Petrick

Acknowledgement

Since the release of the first edition of *The IEEE 802.11 Handbook: A Designer's Companion*, the IEEE 802.11 Working Group has grown to be one of largest wireless standards working groups in the IEEE 802 organization. It has been personally rewarding to us to have worked with some of the finest leading engineering professionals from all over the world in the development of IEEE 802.11™ WLAN standards and amendments for the past 12 years. We would like to thank each of the engineers in the IEEE 802.11 Working Group and the task group chairs who have contributed their work in the creation of the standard's amendments for the medium access control (MAC) sublayer and physical layers (PHYs) covered in this edition of the book.

We thank Tim Godfrey for his contribution on IEEE P802.11e quality of service (QoS) chapter in this book.

We would also like to thank the Wi-Fi Alliance and its membership for their continuing support of the IEEE 802.11 standards. This support ensures certification of interoperable IEEE 802.11 standards-based solutions in the market.

We would like to thank our anonymous reviewers who have provided helpful comments on the refinement of the book.

Finally, we like to thank the IEEE Standards Information Network/IEEE Press and especially thank Jennifer McClain Longman, our editor, for her patience, support, and guidance while completing this book.

Bob O'Hara
Al Petrick

Authors

Bob O'Hara is a co-founder of Airspace, a venture-funded startup company that is leading the IEEE 802.11 industry into the second stage of wireless local area network (WLAN) evolution. He is actively involved in the development of networking, telecommunications, and computing standards and products. His areas of expertise are network and communication protocols and their implementation, operating systems, system specification and integration, standards  development, cryptography and its application, strategy development, and product definition. Mr. O'Hara has been involved with the development of the IEEE 802.11 WLAN standard since 1992. He was the technical editor of that standard and chairman of the revisions and regulatory extensions task groups. He is currently chairman of the maintenance and revision task group. In 2004, he was selected as one of the 50 most powerful people in networking by *Network World*.

Previously, Mr. O'Hara was the president and founder of Informed Technology, Inc., a company that specializes in strategic, technology, and network consulting. Prior to starting Informed Technology, he worked for Advanced Micro Devices in both senior engineering and management positions for the I/O and Network Products Division and in the Advanced Development Lab, as well as in engineering positions at Fairchild Space and Communications and TRW Defense and Space Systems Group. Mr. O'Hara graduated with a BSEE from the University of Maryland in 1978.

Lakshmi,

Thank you for your support, developing great standards!

Bob O'Hara

Al Petrick is vice president of marketing and business development at WiDeFi, a fabless semiconductor company developing 802.11 Wi-Fi® semiconductors for the wireless consumer electronics market. Mr. Petrick's experience includes over 23 years of combined marketing and systems engineering in wireless communications and semiconductor technology. Prior to WiDeFi, he held executive management marketing and business development positions at Icefyre and Intersil. At Intersil, now Conexant,

formally Harris Semiconductor, he pioneered the PRISM WLAN chipset from inception into a successful Wi-Fi product line. Mr. Petrick serves as Vice Chairman of the IEEE 802.11 WLAN Working Group. He has published various marketing and technical papers on wireless communications for leading wireless trade publications, marketing analysts, and financial analysts. Mr. Petrick co-authored with Bob O'Hara the first edition of the *IEEE 802.11 Handbook: A Designer's Companion*. He serves on a number of advisory boards for Wi-Fi product companies. Mr. Petrick holds a BSET from Rochester Institute of Technology, Rochester, New York, and an MBA from Rollins College, Winter Park, Florida, and studied business strategies at Northwestern University's Kellogg Graduate School of Management, Evanston, Illinois.

Lakshmi
Enjoy Reading
the 2nd Edition
Best Wishes!
Al Petr

Contributor

Tim Godfrey is a strategic marketing manager for the PRISM Wireless Business Unit at Conexant. His responsibilities include strategic planning and marketing of broadband wireless products for enterprise and home networking solutions, focusing on the medium access control (MAC) sublayer. Mr. Godfrey has over 25 years of networking and wireless communication experience. His involvement in wireless communication began over 11 years ago when he became director of technology at Digital  Ocean. Mr. Godfrey has been a member of the IEEE 802.11 Working Group since 1994 and is currently the Secretary of that organization. He holds a BSEE ('79) from the University of Kansas.

Foreword

Since the publication of the first edition of the highly successful book, *The IEEE 802.11 Handbook: A Designer's Companion*, the market for the IEEE 802.11 standard and Wi-Fi-enabled products has exploded. Applications provided by equipment manufacturers have entered all segments of our daily life, and I am amazed by the multitude of ideas and functionalities I am seeing in this ever-expanding, worldwide, wireless-enabled push. Key to this explosion is standards-based, interoperable wireless local area network (WLAN) systems.

Since the September 1999 approval by the IEEE-SA Standards Board of the 2.4 GHz, 11 Mbit/s (IEEE 802.11b) and 5 GHz, 54 Mbit/s (IEEE 802.11a) extensions, we have seen further enhancements to the IEEE 802.11 base standard in the areas of domain extensions (IEEE 802.11d), Inter Access Point Protocol (IAPP) (IEEE 802.11F), extended rate 2.4 GHz physical layer (PHY) (IEEE 802.11g), expansion of IEEE 802.11a extension for Europe and Japan (IEEE 802.11h and IEEE 802.11j), and strengthening of the underlying security (IEEE 802.11i).

Further enhancements are to come in the areas of quality of service (QoS), radio resource measurements, fast roaming, extended service set (ESS) mesh networking, wireless performance prediction, advance security, wireless access for vehicular environments, wireless interworking with external networks, wireless network management, and finally higher throughputs for WLAN systems. However, the possibilities are endless, and I am sure the IEEE 802.11 standard will encompass other areas over time, with the help of the drive from the marketplace and the creative knowledge of everyone involved in this exciting time.

Standards are written as specifications for interoperable products and not as handbooks, such as this book, for obtaining a thorough understanding of the protocol and issues involved. It is impossible to include in the standards all the reasons for decisions taken to get the standard ratified, and I would like to congratulate Bob O'Hara and Al Petrick for taking on this enormous task, once again.

I encourage readers of this handbook to be comforted by the very fact that these two authors, as long-time members of the IEEE 802.11 Working Group and standards community, have followed the standards process from the very beginning. They are very able and technically experienced, with a sound understanding of the IEEE 802.11 standard throughout its development stages. This competence is shown throughout this handbook, with clarity of explanation and careful crafting of the content, to bring the reader up to their level of understanding. It is a book with which I am proud to be associated. This handbook is a perfect balance of information for embracing the PHY and medium access control (MAC) sublayer of the entire IEEE 802.11 standard.

I expect this handbook to once again become a standard reference book for every WLAN engineer, student, and integrator and to become the authors' next best seller.

Stuart J. Kerry
Chair, IEEE 802.11, Standards Working Group for Wireless LANs

Philips Semiconductors, Inc.,
1109 McKay Drive, M/S 48A SJ,
San Jose, CA 95131
United States of America

Foreword to First Edition

Since the publication of the IEEE 802.11 WLAN standard, many equipment manufacturers have entered the market with interoperable WLAN systems. In September 1999, the IEEE-SA Standards Board approved the 2.4 GHz, 11 Mbps 802.11b and 5 GHz, 54 Mbps 802.11a extensions. However, standards are written as specifications for interoperable products and not as handbooks for obtaining a thorough understanding of the protocol. It is impossible to include in the standards all the reasons for decisions taken to get the standard ratified.

The only people who could write a handbook with the qualities I have in mind are those that have followed the standards process from the beginning. I applaud Bob O'Hara and Al Petrick for taking on the task of writing this handbook. Bob and Al have been very instrumental throughout the development of the IEEE 802.11 standard and are recognized for their contributions and technical leadership. This book is a first-of-a-kind and provides a perfect balance of information for embracing the physical and MAC layers of the standard.

I expect *The IEEE 802.11 Handbook: A Designer's Companion* to become a standard reference for every WLAN systems engineer and anticipate the reader will find this text extremely useful.

Vic Hayes
Chairman, IEEE P802.11, Standards WG for Wireless LANs

Lucent Technologies Nederland B. V.
Zadelstede 1-10
The Netherlands

Preface

Over five years have passed since publication of the first edition of *The IEEE 802.11 Handbook: A Designer's Companion*. For many people, this seminal book introduced them to the highly complex, yet pivotal, IEEE 802.11 technology known today as *Wi-Fi*. This much-anticipated second edition from Bob O'Hara and Al Petrick is sure to be received enthusiastically by readers as they further yearn to understand important recent enhancements to the underlying standard.

Also during this same period in time, the wireless local area network (WLAN) industry expanded considerably beyond its modest vertical market origins. As a technology that used to be understood only by a handful of information technology (IT) specialists, Wi-Fi has quickly gained rapid acceptance and speedy adoption by millions of nontechnical, home network users. Indeed, Wi-Fi technology is today a very mainstream capability and is featured in an ever-widening array of products. Laptops, set top boxes, televisions, mobile handheld devices, cameras, and countless other products that all employ Wi-Fi connectivity are introduced every week. The principal reason behind this successful market expansion is a common, yet fundamental, ingredient: the underlying worldwide IEEE 802.11 base standard and its related amendments. Further, this common standard provides the basis for strong interoperability certification testing by the Wi-Fi Alliance for all the varieties of products coming to the market.

The IEEE 802.11 base standard was over 400 pages when first published. As this edition of this handbook goes to press, the following amendments have subsequently been ratified: 802.11a, 802.11b, 802.11d, 802.11g, 802.11h, 802.11i, and 802.11j, as well as the recommended practice, 802.11F. And in the works at varying stages of completion are the additional extensions: 802.11e, 802.11k, 802.11m, 802.11n, 802.11p, 802.11r, 802.11s, and 802.11t. As evidence to the brisk pace at which this technology continues to evolve, the IEEE 802.11 Working Group is also poised to authorize its latest task group: 802.11u. The authors, both seasoned experts in the WLAN technology arena and longstanding contributors to IEEE 802.11, guide readers through the complexity of the standard and zero in on the core issues for implementers and adopters alike. In summary, this updated book is the de facto reference that is crucial for anyone developing IEEE 802.11 products or anyone simply wanting to gain a better understanding of the underlying principals of this technology.

Read this with equal doses of curiosity and pleasure.

Bill Carney
Board Chairman
Wi-Fi Alliance (www.wi-fi.org)

Preface to First Edition

This book from Bob and Al is very timely. Wireless LANs are exploding in popularity. The WLAN industry is taking off and expanding beyond its vertical niche market roots. One of the key drivers of this new market expansion for WLANs is the IEEE 802.11 standard. Simply having a WLAN standard was not enough to spark the industry. IEEE 802.11 has been around since June of 1997. The IEEE 802.11b High-Rate Physical Layer extension enabled us to deliver 11 Mbps and products conforming to that standard have been shipping for a while. Wireless LANs have finally hit the right price and performance to appeal to a broader market. Breaking the 10 Mbps barrier makes IEEE 802.11 LANs appealing for enterprise applications. Home networking is becoming more popular, and WLANs are an attractive option. By the time you read this, you will be able to purchase an IEEE 802.11-compliant, 11 Mbps consumer WLAN adapter for $99 or less. Wireless LANs are ready for prime time and IEEE 802.11 made it happen.

The IEEE 802.11 standard represents many years of work from a global team of engineers. One of the challenges of developing the IEEE 802.11 standard was bringing together experts from two different disciplines—analog radio design and network protocol design. We had many arguments about whether this is a radio standard or a network standard. Very clearly, IEEE 802.11 is a network standard. That is the whole point. Because IEEE 802.11 fits into the IEEE 802 framework, systems conforming to the standard can be added to existing networks transparently. IEEE 802.11 WLANs will support the network protocols and applications that were developed for the other IEEE 802 LAN standards over the past 25 years. So IEEE 802.11 is a network standard that happens to have a radio physical layer. This book benefits from the fact that Bob and Al are experts in both of these disciplines. They have a deep understanding of the material gained through their many years of contribution to the standard.

The standard was over 400 pages when initially published, and recently two new physical layer extensions were added. Bob and Al help the reader navigate through the complexity of the standard and focus on the core issues. This book is a great guide to the standard for anyone developing IEEE 802.11 products or those simply wanting to gain a better understanding of the standard.

Enjoy!

Phil Belanger
Chairman of the Wireless Ethernet Compatibility Alliance, www.wi-fi.com
Co-Author of the DFWMAC protocol, the proposal that was used as the basis for the IEEE 802.11 MAC

Table of Contents

Introduction

Quite a lot has happened in the IEEE 802.11™ market and industry since the first edition of *The IEEE 802.11 Handbook: A Designer's Companion* was published at the end of 1999. What was a relatively small market, struggling to expand has blossomed into a robust, growing market, the growth of which outpaced the growth of the general information technology market for several consecutive years. This strong market has spread from largely vertical applications such as healthcare, point of sale, and inventory management to become much more broadly based as a general networking technology being deployed in offices, schools, hotel guest rooms, airport departure areas, airplane cabins, entertainment venues, coffee shops, restaurants, and homes. This expanded demand has led to the growth of new sources of IEEE 802.11 devices, driving the cost of wireless local area networks (WLANs) down dramatically.

In their first generation, WLANs consisted of mobile client devices communicating with independent access points (APs) that concentrated all of the intelligence in each AP. There was little or no cooperation or communication among the APs, which required individual configuration, management, and monitoring. For small WLANs, such as networks in small businesses or homes, designing, installing, and maintaining a WLAN was a task of reasonable complexity. However, as the size of a WLAN and the number of APs in the WLAN increased, the complexity of the task of designing, installing, and maintaining a WLAN grew astronomically. This extraordinary complexity limited the acceptance of the first stage of IEEE 802.11 equipment in large enterprises. The appearance of dedicated gateway devices that added security and mobility functions that were lacking in first-stage devices were the first attempts to address some of the complexity of large WLANs. The gateways would isolate the WLAN and the mobile devices on it from the rest of the network to which the gateway was connected and thereby enforce security policies or enable enhanced mobility in large networks.

As this book is being written, IEEE 802.11 equipment is moving into its second stage, where the WLAN is being treated as a large wireless communication *system*. As a system, there is more to consider than simply the communication over the air between a single AP and the associated mobile devices. This change in the way manufacturers perceive the WLAN has lead to innovative changes in the equipment that makes up a WLAN. One of the most significant changes in WLAN equipment has been brought by the introduction of WLAN *switches*. These switches are devices that centralize the control and management of a WLAN, when deployed in combination with *lightweight* APs.

Centralization, by itself, is not necessarily a significant improvement in a WLAN. WLAN switches and lightweight APs utilized the concept of centralization to introduce a hierarchy into WLANs that was not present before. This hierarchy makes much more information about the WLAN available in a single location than could be available to isolated, independent APs. This information can now be used for automated configuration, coordinated management of the radio domain, load management of the mobile devices, location services for mobile devices, detection and location of rogue APs, and coordinated management of the operating software on all of the lightweight APs in the WLAN.

We have written this second edition of *The IEEE 802.11 Handbook* for the designers of the IEEE 802.11 equipment at the heart of this second stage in the evolution of WLANs and for the designers who will take IEEE 802.11 to the next stage.

Quick guide to the IEEE 802.11 standard

Two major components of the WLAN are described by the IEEE 802.11 standard: the mobile station (STA) and the AP. Going well beyond what other IEEE 802® standards have done in the past, IEEE 802.11 defines a complete management protocol between the mobile STA and AP. This management protocol makes it possible for a single IEEE 802.11 WLAN to comprise equipment from many vendors and marks true multivendor interoperability.

There is a huge amount of information in the IEEE 802.11 standard and its amendments. Finding the required information in a short time can be challenging. To help meet the challenge, a mapping between the information in the standard and the information presented in this handbook is given here. IEEE standards are divided into clauses and annexes, and as is the case with the IEEE 802.11 standard, may have amendments to the base standard.[a] Information in the standard is referred to by the clause and annex in which it is found. This handbook is divided into chapters. Information in this handbook is referred to by the chapter in which it is found.

Clauses 1 through 4 of the standard contain a brief overview of the standard, other references that are required to implement the standard, definitions of terms, and the abbreviations and acronyms used in the standard. This information corresponds to the introduction and abbreviations in this handbook.

Clause 5 of the standard provides a description of the architecture and components of an IEEE 802.11 WLAN system. This information corresponds to Chapter 2 in this handbook.

Clause 6 of the standard describes the medium access control (MAC) service interface. This is an abstract interface for the exchange of data between the MAC and the protocol layer above the MAC. This is not described explicitly in this handbook.

Clause 7 of the standard describes the MAC frames and their content. Clause 8 of the standard was entirely rewritten with IEEE 802.11i and describes security enhancements, including the wired equivalent privacy (WEP) functionality and the robust security network (RSN). Clause 9 describes the functionality and frame exchange protocols of the MAC. Information from these clauses is found in Chapter 3 and Chapter 4 of this handbook.

---

[a] In this handbook, IEEE Std 802.11, 1999 Edition (Reaff 2003), (meaning the base standard and all its amendments) is called "the IEEE 802.11 standard." When distinctions are needed between the base standard and its amendments, IEEE Std 802.11-1999 is called "the IEEE 802.11 base standard," and the amendments use an abbreviated form of their designation, e.g., "IEEE 802.11a" or "IEEE 802.11b."

Clause 10 of the standard describes the layer management service interface primitives and their functionality. Clause 11 describes the MAC management functionality and protocols. This information may be found in Chapter 9 of this handbook.

Clause 12 of the standard describes the physical layer (PHY) service interface. This is an abstract interface for the exchange of data between the MAC and PHY. Clause 13 describes the PHY management service interface, which consists solely of the management information base (MIB) interface. This is not described explicitly in this handbook.

Clause 14 of the standard describes the frequency hopping spread spectrum (FHSS) PHY. Clause 15 describes the direct sequence spread spectrum (DSSS) PHY. Clause 16 describes the infrared baseband PHY. Clause 17 (introduced in IEEE 802.11a and enhanced by IEEE 802.11j) describes the orthogonal frequency division multiplexed (OFDM) PHY. Clause 18 (introduced in IEEE 802.11b) describes the higher rate direct sequence spread spectrum (HR/DSSS) PHY. Information on all PHYs is found in Chapter 11 of this handbook while Chapter 12 and Chapter 13 discuss enhancements to the OFDM PHY and HR/DSSS PHY based on the amendments.

Clause 19 of the standard (introduced in IEEE 802.11g) describes another PHY that allows the use of even higher data rates in the 2.4 GHz frequency band to accommodate the growing demand for wireless connectivity. This clause is discussed in Chapter 14 of this handbook.

Annex A of the standard is the Protocol Implementation Conformance Statement (PICS) proforma. This form may be used to identify the exact options implemented in a device claiming conformance to IEEE 802.11. This annex is not discussed in this handbook.

Annex B of the standard is a set of tables of the hopping patterns for the frequency hopping (FH) PHY. This annex is not discussed in this handbook.

Annex C of the standard is the state machine description of the MAC and MAC management functionality. A discussion of the state machines is beyond the scope of this handbook.

Annex D of the standard is the MIB, written in Abstract Syntax Notation 1 (ASN.1) to comply with the requirements of the Simple Network Management Protocol version 2 (SNMPv2). The MAC portion of the MIB is discussed in Chapter 10 of this handbook.

Annex E of the standard offers a bibliography of related resources.

Annex F of the standard (introduced in IEEE 802.11b) is an informative annex discussing interoperability and clear channel assessment (CCA) recommendations for HR/DSSS PHYs and is not discussed specifically in this handbook.

Annex G of the standard (introduced in IEEE 802.11a) offers an example of frame encoding for an OFDM PHY and is not discussed specifically in this handbook.

Annex H of the standard (introduced in IEEE 802.11i) offers robust security network association (RSNA) reference implementations and test vectors and is not discussed specifically in this handbook.

Annex I and Annex J of the standard (introduced in IEEE 802.11j) offer more information about regulatory classes relative to the Country information element and 4.9-5.0 GHz operation in Japan. These annexes are not discussed specifically in this handbook.

Acronyms and abbreviations

The following acronyms and abbreviations are used in this book:

AC	access category
ACK	acknowledgment frame
ACR	adjacent channel rejection
ADC	analog-to-digital converter
ADDTS	add traffic specification
AES	advanced encryption standard
AGC	automatic gain control
AID	association identifier
AIFS	arbitration interframe space
AKMP	Authentication and Key Management Protocol
AP	access point
APS	asynchronous power save
APSD	automatic power save delivery
ATIM	announcement traffic indication message
AWG	additive white Gaussian
BCC	binary convolutional code
BER	bit error rate
BPSK	binary phase shift keying
BSA	basic service area
BSS	basic service set
BSSID	basic service set identifier
BT	bit period
CBC-MAC	cipher block chaining with message authentication code
CCA	clear channel assessment
CCK	complementary code keying
CCMP	Counter with CBC-MAC Protocol

CDMA	code division multiple access
CF-ACK	contention-free acknowledgment
CFB	contention-free burst
CF-End	contention-free end
CFP	contention-free period
CF-Poll	contention-free poll
CP	contention period
CRC	cyclic redundancy code
CSMA/CA	carrier sense multiple access with collision avoidance
CTS	clear to send
CW	contention window or continuous wave
DA	destination address
DBPSK	differential binary phase shift keying
DCF	distributed coordination function
DFS	dynamic frequency selection
DHCP	Dynamic Host Configuration Protocol
DIFS	distributed interframe space
DLS	direct link setup
DPSK	differential phase shift keying
DQPSK	differential quadrature phase shift keying
DS	distribution system
DSSS	direct sequence spread spectrum
DVT	digital television
EAP	Extensible Authentication Protocol
EAPOL	EAP over LAN
EDCA	enhanced distributed channel access
EIFS	extended interframe space
EIRP	equivalent isotropically radiated power
EOSP	end of service period
ERP	extended-rate PHY
ESS	extended service set

ETSI	European Telecommunications Standards Institute
EVM	error vector magnitude
FCC	Federal Communications Commission
FCS	frame check sequence
FDDI	finer distributed data interference
FFT	Fast Fourier Transform
FH	frequency hopping
FHSS	frequency hopping spread spectrum
GFSK	Gaussian frequency shift key
GPS	global positioning system
GTK	group temporal key
GTKSA	group temporal key security association
HC	hybrid coordinator
HCCA	hybrid coordinated channel access
HCF	hybrid coordination function
HDTV	high-definition television
HMAC	hashed message authentication (or authenticator) code
HR/DSSS	high-rate direct sequence spread spectrum
I/Q	in-phase/quadrature
IAPP	Inter Access Point Protocol
IBSS	independent basic service set
ICI	interchip interference
ICV	integrity check value
IR	infrared
ISI	intersymbol interference
ISM	industrial, scientific, and medical
ITU	International Telecommunications Union
KCK	key confirmation key
KEK	key encryption key
LAN	local area network
LBT	listen before talk

LLC	logical link control
MAC	medium access control
MD-5	Message Digest 5
MIB	management information base
MIC	message integrity check
MIMO	multiple input multiple output
MMACS	Multimedia Mobile Access Communication System
MPDU	MAC protocol data unit
MPHPT	Ministry of Public Management, Home Affairs, Posts and Telecommunications (Japan)
MSDU	MAC service data unit
NAV	network allocation vector
NF	noise figure
NIC	network interface card
OFDM	orthogonal frequency domain multiplexing
OSI	Open System Interconnection
OUI	organizational unique identifier
PBCC	packet binary convolutional code
PC	point coordinator; also, personal computer
PCF	point coordination function
PDA	personal digital assistant
PER	packet error rate
PHY	physical layer
PIFS	priority interframe space
PLCP	physical layer convergence procedure
PMD	physical medium dependent
PMK	pairwise master key
PMKID	pairwise master key identifier (or identity)
PMKSA	pairwise master key security association
PPDU	PLCP protocol data unit
PPM	pulse position modulation

PSDU	PLCP service data unit
PSF	PLCP signaling field
PSK	preshared key
PS-Poll	power save poll
PTK	pairwise transient key
PTKSA	pairwise transient key security association
QAM	quadrature amplitude modulation
QAP	QoS access point
QBSS	QoS BSS
QoS	quality of service
QPSK	quadrature phase shift keying
QSTA	QoS station
RA	receiver address
RADIUS	remote authentication dial-in user service
RF	radio frequency
RFID	radio frequency ID
RLAN	radio local area network
rms	root-mean-square
RSADSI	RSA Data Security, Inc.
RSN	robust security network
RSNA	robust security network association
RTS	request to send
SA	source address
SAP	service access point
SDTV	standard definition television
SFD	start of frame delimiter
SHA-1	Secure Hash Algorithm 1
SIFS	short interframe space
SNMPv2	Simple Network Management Protocol version 2
SNR	signal to noise ratio
SSID	service set identity

STA	station
TA	transmitter address
TBTT	target beacon transmission time
TID	traffic identifier
TIM	traffic indication map
TK	temporal key
TKIP	Temporal Key Integrity Protocol
TPC	transit power control
TSF	timer synchronization function
TSID	traffic stream identifier
TSN	transition security network
TSPEC	traffic specification
TU	time unit
TXOP	transmit opportunity
U-NII	unlicensed national information infrastructure
VLAN	virtual local area network
VoIP	Voice Over Internet Protocol
WEP	wired equivalent privacy
WLAN	wireless LAN

Chapter 1 Similarities and differences between wireless and wired local area networks (LANs)

There are many similarities and differences between wired LANs and the IEEE 802.11™ wireless LAN (WLAN). This chapter will describe them.

SIMILARITIES BETWEEN WLANs AND WIRED LANs

From the beginning, the IEEE 802.11 WLAN was designed to look and feel like any IEEE 802® wired LAN. In other words, it must appear to be the same as the wired networks to which a user may be accustomed. It must support all of the protocols and all of the LAN management tools that operate on a wired network.

To maintain similarity to wired LANs, IEEE 802.11 is designed to the same interface as IEEE 802.3™. IEEE 802.11 operates under the IEEE 802.2™ logical link control (LLC) sublayer, providing all of the services required to support that sublayer. In this fashion, IEEE 802.11 is indistinguishable from IEEE 802.3 by the protocols that may be running above IEEE 802.2.

Using the IEEE 802.2 interface guarantees that protocol layers above LLC need not be aware of the network that is actually transporting their data.

DIFFERENCES BETWEEN WLANs AND WIRED LANs

There are also a number of differences between wired LANs and WLANs. The two most important differences are that there are no wires (the air link) and the mobility thus conferred by the lack of a wired tether. These differences lead to both the tremendous benefits of a WLAN, as well as the perceived drawbacks to them.

The air link is the radio or infrared link between WLAN transmitters and receivers. Because WLAN transmissions are not confined to a wire, there may be concerns that the data carried by a WLAN are not private, i.e., not protected. This concern is certainly valid; the data on a WLAN are broadcast for all to hear. Many proprietary WLANs do not provide any protection for the data carried. The designers of IEEE 802.11 realized that this concern could be a significant problem for users wishing to use a WLAN and designed strong cryptographic mechanisms into the protocol to provide protection for the data that is at least as strong as sending the data over a wire. Details of this protection are described in Chapter 4.

The air link also exposes the transmissions of a WLAN to the vagaries of electromagnetic propagation. For both radio- and infrared-based WLANs, everything in the environment is either a reflector or an attenuator of the signal carrying the LAN data. This variability can cause significant changes in the strength of a signal received by a WLAN station (STA) and sometimes sever the STA from the LAN entirely. At the wavelengths used in the IEEE 802.11 WLAN, small changes in position can cause large changes in the received signal strength. This fluxuation is due to the signal's traveling many paths of differing lengths to arrive at the receiver. Each individual arriving signal is of a slightly different phase from all of the others. Adding these different phases together results in the composite signal that is received. Because these individual signals sometimes add up in phase and sometimes out of phase, the overall received signal strength is sometimes large and sometimes small. Objects moving in the environment, such as people, aluminized Mylar balloons, doors, and other objects, can also affect the strength of a signal at a receiver by changing the attenuation or reflection of the many individual signals.

Figure 1–1 is taken from the IEEE 802.11 standard and shows the result of a ray tracing simulation in a closed office environment. The various shades of gray depict the different signal strengths at each location in the room. Dealing with the variability of the air link is also designed into the IEEE 802.11 WLAN. For more on this, see Chapter 9.

The second significant difference a WLAN has from a wired LAN is mobility. The user of a WLAN is not tethered to the network outlet in the wall. This mobility is both the source of the benefits of a WLAN and the cause of much of the internal complexity.

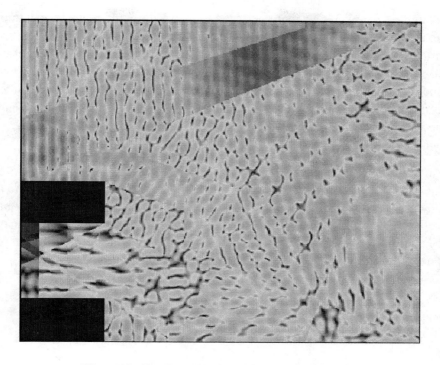

Figure 1–1:   Ray tracing simulation results

The benefit of mobility is that the LAN goes wherever you are, instantly and without the need to search out outlets or to arrange in advance with the network administrators. In a laptop equipped with an IEEE 802.11 WLAN connection, the connection to the network is available in a coworker's office, down the hall in the conference room, downstairs in the lobby, across the parking lot in another building, even across the country on another campus. In other words, all of the information available over the network, while sitting in your office, is still available in all these locations: email, file servers, the company-internal web sites, and the Internet.

Of course, there is a flip side to the benefits of mobility. Most of the network protocols and equipment in use today were not designed to cope with mobility. They were designed with an assumption that the addresses assigned to a network node would remain in a fixed location on the network. For example, early WLANs required that a mobile STA could roam only within an area where the WLAN was connected to the wired LAN, with only layer-2 bridges between the parts of the WLAN. This requirement existed because there was no simple way to deal with the change of a layer-3 network address should the mobile STA cross from one part of the network to another that is connected by a router. Today, there are ways to deal with this problem using new protocols, including Dynamic Host Configuration Protocol (DHCP) and Mobile-IP.

Another problem introduced by mobility is that location-based services lose their "hook" to a user's location when network addresses are not locked to a physical location. Thus, notions such as *the nearest network printer* must be defined in a different way when the physical location of a network user may be constantly changing. This complication may increase the complexity of the service location provider, but meets the needs of the mobile user.

Chapter 2 IEEE 802.11: First international standard for WLANs

In 1997, the IEEE adopted the first standard for WLANs, IEEE Std 802.11. This standard was revised in 1999. The IEEE 802.11 standard defines a medium access control (MAC) sublayer, MAC management protocols and services, and three physical layers (PHYs). The three PHY s are an infrared (IR) baseband PHY, a frequency hopping spread spectrum (FHSS) radio in the 2.4 GHz band, and a direct sequence spread spectrum (DSSS) radio in the 2.4 GHz band. All three PHYs describe both 1 Mbit/s and 2 Mbit/s operation. This chapter will introduce the standard and its concepts.

In 1999, the IEEE 802.11 Working Group standardized two new PHYs. The first, IEEE Std 802.11a, is an orthogonal frequency division multiplexing (OFDM) radio in the U-NII bands, delivering up to 54 Mbit/s data rates. The second, IEEE Std 802.11b, is an extension to the DSSS PHY in the 2.4 GHz band, delivering up to 11 Mbit/s data rates. In 2002, the working group completed the standardization of an extension to 802.11b, IEEE Std 802.11g, that adds all of the OFDM capabilities to radios operating in the 2.4 GHz band.

The goal of the IEEE 802.11 standard is to describe a WLAN that delivers services previously found only in wired networks, e.g., high throughput, highly reliable data delivery, and continuous network connections. In addition, IEEE 802.11 describes a WLAN that allows transparent mobility and built-in power saving operations to the network user. The remainder of this chapter will describe the architecture of the IEEE 802.11 network and the concepts that support that architecture.

IEEE 802.11 ARCHITECTURE

The architecture of the IEEE 802.11 WLAN is designed to support a network where most decision making is distributed to the mobile STAs. This architecture has several advantages, including being very tolerant of faults in all of the WLAN equipment and eliminating any possible bottlenecks a centralized architecture would introduce. The architecture is very flexible, easily supporting both small, transient networks and large semipermanent or permanent networks. In addition, deep power-saving modes of operation are built into the architecture and protocols to prolong the battery life of mobile equipment without losing network connectivity. The IEEE 802.11 architecture comprises several components: the STA, the access point (AP), the wireless medium, the basic service set (BSS), the distribution system (DS), and the extended service set (ESS). The architecture also includes STA services and distribution services.

The IEEE 802.11 architecture may appear to be overly complex. However, this apparent complexity is what provides the IEEE 802.11 WLAN with its robustness and flexibility. The architecture also embeds a level of indirection that has not been present in previous LANs. It is this level of indirection, handled entirely with the IEEE 802.11 architecture and transparent to protocol users of the IEEE 802.11 WLAN, that provides the ability of a mobile STA to roam throughout a WLAN and appear to be stationary to the protocols above the MAC that have no concept of mobility. This "sleight of hand" performed by IEEE 802.11 allows all of the existing network protocols to run over a WLAN without any special considerations.

STA

The STA is the component that connects to the wireless medium. It consists of a MAC and a PHY. Generally, the STA may be referred to as the *network adapter* or *network interface card (NIC)*. These names may be more familiar to users of wired networks.

The STA may be mobile, portable, or stationary. Every STA supports STA services. These services are authentication, deauthentication, privacy, and delivery of the data. The STA services are described in this chapter.

Basic service set (BSS)

The IEEE 802.11 WLAN architecture is built around a BSS, which is a set of STAs that communicate with one another. A BSS does not generally refer to a particular area, due to the uncertainties of wireless propagation. When all of the STAs in the BSS are mobile STAs and there is no relay connection to a wired network, the BSS is called an *independent BSS (IBSS)*. The IBSS is the entire network, and only STAs communicating with each other in the IBSS are part of the LAN. An IBSS is typically a short-lived network, with a small number of STAs, that is created for a particular purpose, e.g., to exchange data with a vendor in the lobby of your company's building or to collaborate on a presentation at a conference.

In an IBSS, the mobile STAs all communicate directly with one another. Not every mobile STA may be able to communicate with every other mobile STA, but they are all part of the same IBSS. There is also no relay function in an IBSS. Thus, if one mobile STA must communicate with another, the two STAs must be in direct communication range. See Figure 2–1.

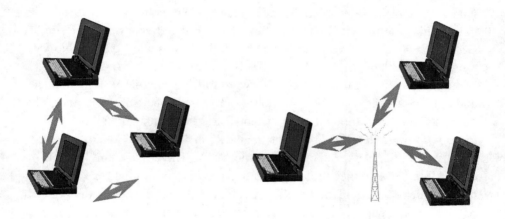

Figure 2–1:   IBSS

When a BSS includes an AP, the BSS is no longer independent. It is called an *infrastructure BSS*, but often referred to simply as a BSS. An AP is a STA that also provides distribution services. Distribution services are described later in this chapter.

In an infrastructure BSS, all mobile STAs communicate with the AP. The AP provides both the connection to the wired LAN, if any, and the local relay function for the BSS. Thus, if one mobile STA in the BSS must communicate with another mobile STA, the communication is sent first to the AP and then from the AP to the other mobile STA. This process causes communications that both originate and end in the same BSS to consume twice the bandwidth that the same communication would consume if sent directly from one mobile STA to another. While this consumption appears to be a significant cost, the benefits provided by the AP far outweigh this cost. One of the benefits provided by the AP is the buffering of traffic for a mobile STA while that STA is operating in a very low power state. The protocols and mechanisms for the support of power saving by mobile STAs are described in Chapter 9.

Extended service set (ESS)

One of the most desirable benefits of a WLAN is the mobility it provides to its users. This mobility would not be of much use if it were confined to a single BSS. IEEE 802.11 extends the range of mobility it provides to any arbitrary range through the ESS. An ESS is a set of infrastructure BSSs, where the APs communicate among themselves to forward traffic from one BSS to another and to facilitate the movement of mobile STAs from one BSS to another. The APs perform this communication via an abstract medium called the *distribution system (DS)*. The DS is the backbone of the WLAN and may be constructed of either wired or wireless networks. The DS is a thin layer in each AP that determines if communications received from the BSS are to be relayed back to a destination in the BSS, forwarded on the DS to another AP, or sent into the wired network infrastructure to a destination not in the ESS. Communications received by an AP from the DS are transmitted to the BSS to be received by the destination mobile STA. To network equipment outside of the ESS, the ESS and all of its mobile STAs appears to be a single MAC sublayer network where all STAs are physically stationary. Thus, the ESS hides the mobility of the mobile STAs from everything outside the ESS. This level of indirection, provided by the IEEE 802.11 architecture, allows existing network protocols that have no concept of mobility to operate correctly with a WLAN where there is lots of mobility. See Figure 2–2.

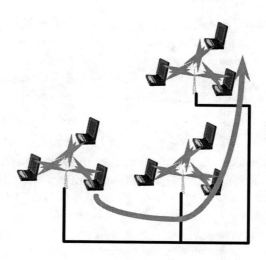

Figure 2–2:   ESS

One area that is beyond the scope of the IEEE 802.11 standard itself is the
communication between APs. The 802.11 Working Group completed work on
a recommended practice for the Inter Access Point Protocol (IAPP). IEEE
Std 802.11F describes the functionality and application of the protocol and is
discussed in Chapter 8.

Distribution system (DS)

The DS is the mechanism by which one AP communicates with another to
exchange frames for STAs in their BSSs, forward frames to follow mobile
STAs from one BSS to another, and exchange frames with wired networks, if
any. As IEEE 802.11 describes it, the DS is not necessarily a network. The
standard does not place any restrictions on how the DS is implemented, only
on the services it must provide. Thus, the DS may be a wired network, such as
IEEE 802.3, or it may be a purpose-built box that interconnects the APs and
provides the required distribution services. These services are described in
this chapter.

SERVICES

There are nine services defined by the IEEE 802.11 architecture. These services are divided into two groups: STA services and distribution services. The STA services comprise authentication, deauthentication, privacy, and delivery of the data. The distribution services comprise association, reassociation, disassociation, distribution, and integration.

STA services

The four STA services—authentication, deauthentication, privacy, and data delivery—provide the IEEE 802.11 WLAN similar functions to the functions that are expected of a wired network. The wired network function of physically connecting to the network cable is similar to the authentication and deauthentication services, where use of the network is allowed only to authorized users. The authentication service is used to prove the identity of one STA to another. Without this proof of identity, the STA is not allowed to use the WLAN for data delivery. The deauthentication service is used to eliminate a previously authorized user from any further use of the network. Thus, once a STA is deauthenticated, e.g., when an employee resigns, that STA can no longer access the services of the IEEE 802.11 WLAN.

The privacy service of IEEE 802.11 is designed to provide a level of protection for data traversing the WLAN equivalent to the level provided by a wired network that exists in an office building with restricted physical access to the network plant. This service protects the data only as they traverse the wireless medium. It is not designed to provide complete protection of data between applications running over a mixed network environment that happens to include an IEEE 802.11 WLAN.

Finally, the data delivery service of an IEEE 802.11 WLAN is similar to the service provided by all other IEEE 802 LANs. The data delivery service provides reliable delivery of data frames from the MAC in one STA to the MAC in one or more other STAs, with minimal duplication and minimal reordering.

Distribution services

The five distribution services—association, reassociation, disassociation, distribution, and integration— provide the services necessary to allow mobile STAs to roam freely within an ESS and allow an IEEE 802.11 WLAN to connect with the wired LAN infrastructure. The distribution services operate in a thin layer above the MAC and below the LLC sublayer and are invoked to determine how to forward frames within the IEEE 802.11 WLAN and also how to deliver frames from the IEEE 802.11 WLAN to network destinations outside of the WLAN.

The *association service* is used to make a logical connection between a mobile STA and an AP. This logical connection is necessary in order for the DS to know where and how to deliver data to the mobile STA. The logical connection is also necessary for the AP to accept data frames from the mobile STA and to allocate resources to support the mobile STA. Typically, the association service is invoked once, when the mobile STA enters the WLAN for the first time (i.e., after the application of power or when rediscovering the WLAN after being out of touch for a time).

The *reassociation service* is similar to the association service, with the exception that it includes information about the AP with which a mobile STA has been previously associated. A mobile STA will use the reassociation service repeatedly as it moves throughout the ESS, loses contact with the AP with which it is associated, and needs to become associated with a new AP. By using the reassociation service, a mobile STA provides information to the AP to which it will be associated that allows that AP to contact the AP with which the mobile STA was previously associated, to obtain frames that may be waiting there for delivery to the mobile STA, as well as other information that may be relevant.

The *disassociation service* is used either by an AP to force a mobile STA to associate elsewhere or by a mobile STA to inform an AP that it no longer requires the services of the WLAN. An AP may use the disassociation service to inform one or more mobile STAs that the AP can no longer provide the logical connection to the WLAN. This development may be due to demand exceeding available resources in the AP, the AP's shutting down, or any number of other reasons. When the mobile STA becomes disassociated, it must begin a new association by invoking the association service.

A mobile STA may also use the disassociation service. When a mobile STA is aware that it will no longer require the services of the AP, it may invoke the disassociation service to notify the AP that the logical connection to the WLAN from this mobile STA is no longer required. For example, this notification may be done when the mobile STA is being shut down or when the IEEE 802.11 adapter card is being ejected. At that point, an AP may free any resources dedicated to the mobile STA and recover them for other uses.

An AP uses the *distribution service* to determine how to deliver the frames it receives. When a mobile STA sends a frame to the AP for delivery to another STA, the AP invokes the distribution service to determine if the frame should be sent back into its own BSS, for delivery to a mobile STA that is associated with the AP, or if the frame should be sent into the DS for delivery to another mobile STA associated with a different AP or to a network destination outside the IEEE 802.11 WLAN. The distribution service determines if the frame is sent to another AP or to a portal.

The *integration service* connects the IEEE 802.11 WLAN to other LANs, including one or more wired LANs, or other IEEE 802.11 WLANs. A portal performs the integration service. The portal is an abstract architectural concept and may physically reside as a thin layer in some or all APs, or it may be a separate network component entirely. The integration service translates IEEE 802.11 frames to frames that may traverse another network and, vice versa, translates frames from other networks to frames that may be delivered by an IEEE 802.11 WLAN.

Interaction between some services

The IEEE 802.11 standard states that each STA must maintain two variables that are dependent on the authentication/deauthentication services and the association/reassociation/disassociation services. The two variables are authentication state and association state. While the standard describes these variables as being enumerated types, they are available only internal to an implementation and can be implemented as Boolean truth-values. The variables are used in a simple state machine that determines the order in which certain services must be invoked and the time at which a STA may begin using the data delivery service. The variables must exist in enough instances to allow the STA to maintain a unique copy for each STA with which it communicates. A STA may be authenticated with many different STAs simultaneously. However, a STA may be associated with only one other STA at a time.

A STA begins operation in state 1, where both authentication state and association state are false to indicate that the STA is neither authenticated nor associated. In state 1, a STA may use a very limited number of frame types. (The details of the frame types are described in Chapter 3.) The allowable frame types provide the capability for a STA in state 1 to find an IEEE 802.11 WLAN, an ESS, and its APs; to complete the required frame handshake protocols; and to implement the authentication service. If a STA is not successful in becoming authenticated, it will remain in state 1. If a STA becomes authenticated, authentication state is set to true, and the STA will make a transition to state 2. See Figure 2–3.

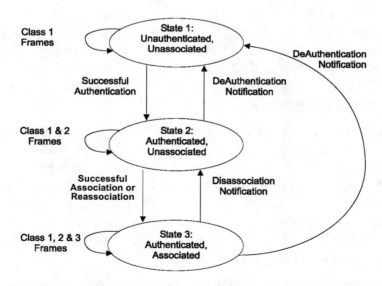

Figure 2–3: Relationship between state variables and services

If a STA is part of an IBSS, it is allowed to implement the data service in state 1. This capability is allowed because neither authentication nor association is used in an IBSS, leaving no mechanism for a STA in an IBSS to leave state 1.

In state 2, the STA has been authenticated (i.e., authentication state is true), but not yet associated. In this state, additional frame types are allowed, beyond the frame types allowed in state 1. The additional frame types provide the capability for a STA in state 2 to implement the association, reassociation, and disassociation services. If a STA becomes associated, association state is set to true, and the STA will make a transition to state 3.

If a STA is not successful in becoming associated, it will remain in state 2. If it receives a deauthentication notification while in state 2, it will return to state 1, and authentication state will be made false.

In state 3, the STA has been both authenticated and associated (i.e., both authentication state and association state are true). In this state, all frame types are allowed, and the STA may use the data delivery service. A STA will remain in this state until it receives either a disassociation notification or a deauthentication notification or until it reassociates with another STA. If a STA receives a disassociation notification, it will make a transition to state 2 and set association state to false. If a STA receives a deauthentication notification, it will make a transition to state 1 and set both authentication state and association states to false.

A STA must react to frames it receives in each of the states, even frames that are disallowed for a particular state. A STA will send a deauthentication notification to any STA with which it is not authenticated if it receives frames that are not allowed in state 1. A STA will send a disassociation notification to any STA with which it is authenticated, but not associated, if it receives frames not allowed in state 2. These notifications will force the STA that sent the disallowed frames to make a transition to the proper state in the state diagram and allow it to proceed properly toward state 3.

It can now be seen that a STA will make transitions between the states of this state machine many times as it roams through an ESS. Because a STA may be authenticated with many STAs at once, it may be in state 2 with relation to those STAs. However, a STA may be in state 3 with relation to only a single other STA. When a STA reassociates with another STA, the STA with which it was previously associated must be moved back to state 2, by setting the value of associated state for that STA to false.

As a graphical example of how the services are used, Figure 2–4 shows a STA moving between APs. As the STA finds AP1, it will authenticate and associate (a). As the STA moves, it may preauthenticate with AP2 (b). When the STA determines that its association with AP1 is no longer desirable, it may reassociate with AP2 (c). The reassociation causes AP2 to notify AP1 of the new location of the STA so the STA's previous association with AP1 can be terminated (d). At this point, the STA would need to find another AP with which to authenticate and associate, in order to continue using the WLAN (f).

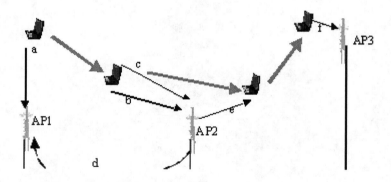

Figure 2–4:   Sample usage of services

SUMMARY

The architecture and services of IEEE 802.11 are designed to allow the WLAN to appear identical to wired LANs. The architecture clearly divides the functionality of the WLAN into nonoverlapping functional blocks. The services described by the IEEE 802.11 standard provide the user of IEEE 802.11 with the functionality of a wired LAN and the additional benefits of nearly ubiquitous mobility.

Chapter 3 Medium access control (MAC)

The IEEE 802.11 MAC is the sublayer that supplies the functionality required to provide a reliable delivery mechanism for user data over noisy, unreliable wireless media. In addition, it provides advanced LAN services, equal to or beyond the services of existing wired LANs.

MAC FUNCTIONALITY

The first function of the MAC is to provide a reliable data delivery service to the users of the MAC. Through a frame exchange protocol at the MAC level, the IEEE 802.11 MAC significantly improves on the reliability of data delivery over wireless media, as compared to earlier WLANs.

The second function of the IEEE 802.11 MAC is to fairly control access to the shared wireless medium. It performs this function through two different access mechanisms: the basic access mechanism, called the *distributed coordination function* (DCF), and a centrally controlled access mechanism, called the *point coordination function* (PCF).

The third function of the IEEE 802.11 MAC is to protect the data that it delivers. Because a WLAN cannot be contained to a particular physical area, in most cases, the IEEE 802.11 MAC provides a privacy service that encrypts the data sent over the wireless medium. The level of encryption chosen approximates the level of protection data might have on a wired LAN in a building with controlled access that prevents any physical connections to the LAN wiring without authorization.

MAC FRAME EXCHANGE PROTOCOL

Because the media used by the IEEE 802.11 WLAN are often very noisy and unreliable, the IEEE 802.11 MAC implements a frame exchange protocol to allow the source of a frame to determine when the frame has been successfully received at the destination. This frame exchange protocol adds some overhead beyond that of other MAC protocols, like IEEE 802.3,

---

because to simply transmit a frame and expect that the destination has received it correctly is not sufficient on wireless media. In addition, we cannot expect that every STA in a WLAN is able to communicate directly with every other STA in the WLAN. This limitation leads to a situation called the *hidden node problem*. The MAC frame exchange protocol is also designed to address this problem of WLANs. The frame exchange protocol requires the participation of all STAs in the WLAN. For this reason, every STA decodes and reacts to information in the MAC header of every frame it receives.

Dealing with the media

The minimal MAC frame exchange protocol consists of two frames: a frame sent from the source to the destination and an acknowledgment sent from the destination to confirm that the frame was received correctly. The frame and its acknowledgment are an atomic unit of the MAC protocol. As such, they cannot be interrupted by a transmission from any other STA.

If the source does not receive the acknowledgment (because the destination did not send one due to errors in the original frame or because the acknowledgment itself was corrupted), the source will attempt to transmit the frame again, according to the rules of the basic access mechanism described below. This retransmission of frames by the source effectively reduces the inherent error rate of the medium, at the cost of additional bandwidth consumption. Without this mechanism for retransmission, the users of the MAC, i.e., higher layer protocols, would be left to determine that their packets had been lost through higher layer timeouts or other means. Because higher layer timeouts are often measured in seconds, it is much more efficient to deal with this issue at the MAC sublayer.

Hidden node problem

A WLAN suffers from a problem that does not occur on a wired LAN. This problem is one of "hidden nodes." It is a result of the fact that every WLAN STA cannot be expected to communicate directly with every other WLAN STA. An example illustrates this problem most clearly.

In this example, there are three STAs, A, B, and C, arranged as shown in Figure 3–1. STA A can communicate only with STA B. STA B can communicate with STAs A and C. STA C can communicate only with STA B. If a simple "transmit and hope" protocol were to be used when STA A was sending a frame to STA B, the frame could be corrupted by a transmission begun by STA C. STA C would be completely unaware of the ongoing transmission from STA A to STA B.

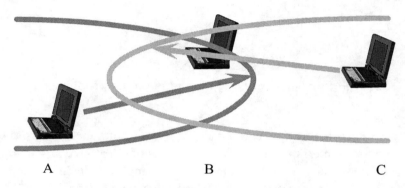

A   B   C

Figure 3–1:   Hidden node problem

The IEEE 802.11 MAC frame exchange protocol addresses this problem by adding two additional frames to the minimal frame exchange protocol described so far. The two frames are a request to send (RTS) frame and a clear to send (CTS) frame. The source sends a RTS frame to the destination. The destination returns a CTS frame to the source. Each of these frames contains information that allows other STAs receiving them to be notified of the upcoming frame transmission and to delay any transmissions of their own. The RTS and CTS frames serve to announce to all STAs in the neighborhood of both the source and destination the impending transmission from the source to the destination. When the source receives the CTS frame from the destination, the real frame that the source wants to deliver to the destination is sent. If that frame is correctly received at the destination, the destination will return an acknowledgment, completing the frame exchange protocol. Depending on the configuration of a STA and its determination of local conditions, a STA may choose when to use the RTS and CTS frames. See Figure 3–2.

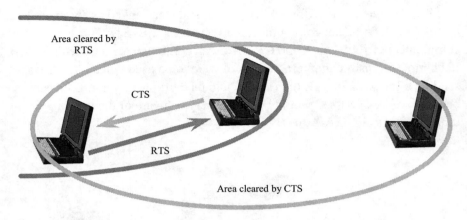

Figure 3–2:   Solving the hidden node problem

The four frames in this exchange are also an atomic unit of the MAC protocol. They cannot be interrupted by the transmissions of other STAs. If this frame exchange fails at any point, the state of the exchange and the information carried in each of the frames allow the STAs that have received these frames to recover and regain control of the medium in a minimal amount of time. A STA in the neighborhood of the source STA receiving the RTS frame will delay any transmissions of its own until it receives the frame announced by the RTS frame. If the announced frame is not detected, the STA may use the medium. Similarly, a STA in the neighborhood of the destination STA receiving the CTS frame will delay any transmissions of its own until it receives the acknowledgment. If the acknowledgment is not detected, the STA may use the medium.

In the source STA, a failure of the frame exchange protocol causes the frame to be retransmitted. This situation is treated as a collision, and the rules for scheduling the retransmission are described in the section on the basic access mechanism, below. To prevent the MAC from being monopolized by attempts to deliver a single frame, there are retry counters and timers to limit the lifetime of a frame.

While this four-way frame exchange protocol is a required function of the MAC, it may be disabled by an attribute in the management information base

(MIB). The value of the dot11RTSThreshold attribute defines the length of a frame that is required to be preceded by the RTS and CTS frames. All frames of a length greater than the threshold will be sent with the four-way frame exchange. Frames of length less than or equal to the threshold will not be preceded by the RTS and CTS frames. This attribute allows a network designer to tune the operation of the IEEE 802.11 WLAN for the particular environment in which it is deployed. In an environment with low demand for bandwidth or where the STAs are concentrated in an area where all are able to hear the transmissions of every STA, the threshold may be set so that the RTS and CTS frames are never used. This setting is the default for the threshold. As long as STAs are not contending with each other, the RTS and CTS frames will most often be consuming bandwidth for no measurable gain. In an environment where there is a significant demand for the bandwidth available in the WLAN or where the STAs are distributed so that some may not hear the transmission of others, the threshold may be set lower, and long frames will have to use the RTS and CTS frame exchange. The value to which the threshold should be set is determined by comparing the bandwidth lost due to the additional overhead of the protocol to the bandwidth lost from transmissions being corrupted by hidden nodes. A typical value for the threshold is 128. However, the value chosen is dependent on the data rate and should be calculated for the particular data rate in use. It is rarely necessary to change the value of the dot11RTSThresold attribute from the default value in an AP. By definition, an AP is heard by all STAs in its BSS and will never be a hidden node. The only situation that may warrant changing the value for the RTS threshold in an AP is when APs are colocated and sharing a channel.

Retry counters

Two retry counters are associated with every frame the MAC attempts to transmit: a short retry counter and a long retry counter. A lifetime timer is also associated with every frame the MAC attempts to transmit. Between these counters and the timer, the MAC may determine that it is no longer worthwhile to continue attempting to transmit a particular frame. When the MAC makes that determination, it may cancel the frame's transmission and discard the frame. If a frame's transmission is canceled, the MAC indicates this to the MAC user through the MAC service interface.

The retry counters limit the number of times a single frame may be retransmitted. There are two counters so that the network designer may choose to allow more or fewer retries to shorter frames, as compared to longer frames. The definition of the length of the frame that uses the short or long counters is determined from the value of the dot11RTSThreshold attribute in the MIB. Upon the initial attempt at the transmission of a frame, the retry counters are reset to zero. Frames of a length less than or equal to this threshold will cause the short retry counter to increment when they are retransmitted. Frames of a length greater than the threshold will cause the long retry counter to increment when they are retransmitted. The successful transmission of a frame will reset the associated retry counter to zero. The attempt to deliver a frame is abandoned if either retry counter reaches the limit with which it is associated in the MIB. The short retry counter is associated with the dot11ShortRetryLimit attribute. The long retry counter is associated with the dot11LongRetryLimit attribute.

BASIC ACCESS MECHANISM

The basic access mechanism is carrier sense multiple access with collision avoidance (CSMA/CA) with binary exponential backoff. This access mechanism is similar to the access mechanism used in IEEE 802.3, with some significant exceptions. CSMA/CA is a "listen before talk" (LBT) access mechanism. In this type of access mechanism, a STA will listen to the medium before beginning a transmission. If the medium is already carrying a transmission, the STA that is listening will not begin its own transmission. This feature is the CSMA portion of the access mechanism and is implemented, in part, using a physical carrier sensing mechanism provided by the PHY. Had the listening STA begun its transmission while the medium was already carrying another transmission, a collision would occur on the medium. The collision may cause one or both of the transmissions to be corrupted to the extent that the transmissions could not be correctly received. Thus, the operation of the access mechanism works to ensure the correct reception of the information transmitted on the wireless medium.

With the IEEE 802.11 access mechanism, when a STA listens to the medium before beginning its own transmission and detects an existing transmission in progress, the listening STA enters a deferral period determined by the binary exponential backoff algorithm. It will also increment the appropriate retry counter associated with the frame. The binary exponential backoff algorithm chooses a random number, which represents the amount of time that must elapse while there are no transmissions, i.e., the medium is idle before the listening STA may attempt to begin its transmission again. The random number resulting from this algorithm is uniformly distributed in a range, called the *contention window* (CW), the size of which doubles with every attempt to transmit a frame that is deferred, until a maximum size is reached for the range. Once a frame is successfully transmitted, the range is reduced to its minimum value for the next transmission. Both the minimum and maximum values for the CW range are fixed for a particular PHY. However, the values may differ from one PHY to another.

Because it is extremely unusual for a wireless device to be able to receive and transmit simultaneously, the IEEE 802.11 MAC uses collision avoidance rather than the collision detection of IEEE 802.3. It is also unusual for all wireless devices in a LAN to be able to communicate directly with all other devices. For this reason, the IEEE 802.11 MAC implements a network allocation vector (NAV). The NAV is a value that indicates to a STA the amount of time that remains before the medium will become available. The NAV is kept current through duration values that are transmitted in all frames. By examining the NAV, a STA may avoid transmitting, even when the medium does not appear to be carrying a transmission by the physical carrier sensing mechanism. The NAV, then, is a virtual carrier sensing mechanism. By combining the virtual carrier sensing mechanism with the physical carrier sensing mechanism, the MAC implements the collision avoidance portion of the CSMA/CA access mechanism.

Timing intervals

The decision by a STA that the medium is not carrying a transmission when the STA is listening before beginning its own transmission is based on timing intervals. The IEEE 802.11 MAC recognizes five timing intervals. Two basic intervals are determined by the PHY: the short interframe space (SIFS) and the slot time. Three additional intervals are built from the two basic intervals: the priority interframe space (PIFS), the distributed interframe space (DIFS), and the extended interframe space (EIFS). The SIFS is the shortest interval, followed by the slot time, which is slightly longer. The PIFS is equal to SIFS plus one slot time. The DIFS is equal to the SIFS plus two slot times. The EIFS is much larger than any of the other intervals. It is used when a frame that contains errors is received by the MAC and allows the possibility for the MAC frame exchanges to complete correctly before another transmission is allowed. Through these five timing intervals, both the DCF and PCF are implemented.

Distributed coordination function (DCF)

The DCF operates as follows: When the MAC receives a request to transmit a frame, a check is made of the physical and virtual carrier sensing mechanisms. If both mechanisms indicate that the medium is not in use for an interval of DIFS (or EIFS if the previously received frame contained errors), the MAC may begin transmission of the frame. If either the physical or virtual carrier sensing mechanisms indicate that the medium is in use during the DIFS interval, the MAC will select a backoff interval using the binary exponential backoff algorithm and increment the appropriate retry counter. The MAC will decrement the backoff value each time the medium is detected to be idle by both the physical and virtual carrier sensing mechanisms for an interval of one slot time. Once the backoff interval has expired, the MAC begins the transmission. It the transmission is not successful, i.e., the acknowledgment is not received, a collision is considered to have occurred. In this case, the CW is doubled, a new backoff interval is selected, and the backoff countdown is begun again. This process will continue until the transmission is sent successfully or it is canceled. See Figure 3–3.

CENTRALLY CONTROLLED ACCESS MECHANISM (I.E., PCF)

The centrally controlled access mechanism uses a poll and response protocol to eliminate the possibility of contention for the medium. This access mechanism is called the *point coordination function (PCF)*. A point coordinator (PC) controls the PCF. The PC is always located in an AP. Generally, the PCF operates as follows: The STAs request that the PC register them on a polling list, and the PC then regularly polls the STAs for traffic

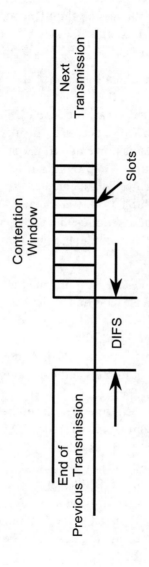

Figure 3–3: DCF timing

while also delivering traffic to the STAs. With proper planning, the PCF is able to deliver a near-isochronous service to the STAs on the polling list. The PCF is built over the DCF, and both operate simultaneously. While the PCF is an optional part of the IEEE 802.11 standard, every STA is required to be able to respond to the operation of the PCF.

The PCF makes use of the PIFS, which is shorter than the DIFS, to seize and maintain control of the medium. The PC begins a period of operation called the *contention-free period* (CFP), during which the PCF is operating. This period is called contention free because access to the medium is completely controlled by the PC and the DCF is prevented from gaining access to the medium. The CFP occurs periodically to provide a near-isochronous service to the STAs. The CFP also alternates with a contention period where the normal DCF rules operate and all STAs may compete for access to the medium. The IEEE 802.11 standard requires that the contention period be long enough to contain at least one maximum length frame and its acknowledgment.

The CFP begins when the PC gains access to the medium, using the normal DCF procedures, and transmits a Beacon frame (described later in this chapter). Beacon frames are required to be transmitted periodically also. Because the PC must compete for the medium, the beginning of the CFP may be delayed from its ideal start time. Once the PC has control of the medium, it begins to deliver traffic to STAs in its BSS and may poll STAs that have requested contention-free service for them to deliver traffic to the PC. Thus, the traffic in the CFP will consist of frames sent from the PC to one or more STAs, followed by the acknowledgment from those STAs. Every IEEE 802.11 STA is capable of receiving frames addressed to it from the PC during the CFP and returning an acknowledgment. In addition, the PC sends a contention-free poll (CF-Poll) frame to STAs that have requested contention-free service. If the STA polled has traffic to send, it may transmit one frame for each CF-Poll frame received. If the STA does not have traffic to send, it does not respond to the poll. The ability to respond to the CF-Poll frame is an option in the IEEE 802.11 standard.

One misconception about the PCF, due to an ambiguity in the IEEE 802.11 standard, is that a STA addressed by a CF-Poll frame can respond to the poll by sending a frame to any other STA, not only to the AP. This operation was contemplated during the development of the standard and subsequently rejected, because of the difficulties that had to be overcome to manage such operations. Unfortunately, the text describing this operation was not entirely removed from the standard.

In order to make the use of the medium more efficient during the CFP, it is possible to piggyback both the acknowledgment and the CF-Poll frame onto data frames. Thus, a frame sent to a STA from the PC may include a CF-Poll frame of that STA for it to send traffic back to the PC. The frame sent from the STA to the PC may include the acknowledgment of the frame just received from the PC. The PC may combine both the CF-Poll and the acknowledgment with a data frame as well. In this last case, the PC may be sending a frame to one STA, along with a CF-Poll frame, and acknowledging a frame received from an entirely different STA.

During the CFP, the PC ensures that the interval between frames on the medium is no longer than PIFS. This step is taken to prevent a STA operating under the DCF from gaining access to the medium. The PC will send a frame to a STA and expect the responding frame, either an acknowledgment or a data frame in response to a CF-Poll, within a SIFS interval. If the response is not received before that SIFS interval expires, the PC will transmit its next frame before a PIFS interval expires after the previous transmission. This activity will continue until the CFP is concluded. This operation of the PC is actually a secondary mechanism to prevent access by the STAs until the CFP is concluded. Another mechanism also keeps the STAs off the medium during the CFP.

The primary mechanism used to prevent STAs from accessing the medium during the CFP is the NAV. The Beacon frame that is sent by the PC at the beginning of the CFP contains information from the PC about the maximum expected length of the CFP. Every STA receiving the Beacon frame will enter this information into its NAV and thus be prevented from independently accessing the medium until the CFP concludes. The use of the PIFS interval, described in the previous paragraph, is a backup mechanism used to prevent STAs that did not receive the Beacon frame from accessing the medium.

Because the CFP is not a true isochronous service, where the total bandwidth demand and thus the time of transmission is known precisely in advance, the PC announces the end of the CFP by transmitting a contention-free end (CF-End) frame. This frame is the formal conclusion of the CFP. It causes the STAs that had set their NAVs from the initial Beacon frame to reset their NAVs. Once the NAVs are reset, the STAs are able to begin the operation of the DCF, independently competing for access to the medium. Like the acknowledgment and CF-Poll frame during the CFP, the CF-End frame may also be combined with an acknowledgment of a data transmission from a mobile STA. See Figure 3–4.

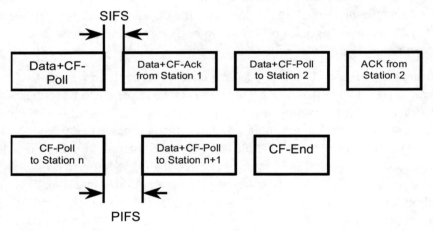

Figure 3–4:   PCF timing

FRAME FORMATS

The IEEE 802.11 MAC accepts data as MAC service data units (MSDUs) from higher layers in the protocol stack for the purpose of reliably sending those MSDUs to the equivalent layer of the protocol stack in another STA. To accomplish this task, the MAC adds information to the MSDU in the form of headers and trailers to create a MAC protocol data unit (MPDU). The MPDU is then passed to the PHY to be sent over the wireless medium to the other STAs. In addition, the MAC may fragment MSDUs into several frames, increasing the probability that each individual frame will be delivered successfully. A discussion of fragmentation follows the description of frame formats in this chapter.

The header and trailer information, combined with the information received as the MSDU, is referred to as the *MAC frame*. This frame contains, among other things, addressing information, IEEE 802.11-specific protocol information, information for setting the NAV, and a frame check sequence (FCS) for verifying the integrity of the frame. The details of the frame format are presented in the following sections.

GENERAL FRAME FORMAT

The general IEEE 802.11 frame format is shown in Figure 3–5. This frame format is more complex than the format for most other LAN protocols. During the development of the IEEE 802.11 standard, a lot of discussion surrounded the frame format. The format that resulted is considered the best design balancing both efficiency and functionality.

The frame begins with a MAC header. The start of the header is the Frame Control field. A field that contains the duration information for the NAV or a short identifier (ID) follows it. Three addressing fields follow that field. The next field contains frame sequence information. The final field of the MAC header is the fourth address field. It appears that the MAC header is very long; however, not all of these fields are used in all frames.

Following the MAC header is the frame body. The frame body contains the MSDU from the higher layer protocols. The final field in the MAC frame is the FCS.

Each of these fields is described in this chapter. Following the field descriptions, the use of the fields in particular frame types is discussed.

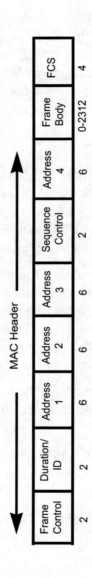

Figure 3–5: IEEE 802.11 frame format

Frame Control field

The Frame Control field is a 16 bits long. It contains the information the MAC requires to interpret all of the subsequent fields of the MAC header.

The subfields of the Frame Control field are Protocol Version, [frame] Type, Subtype, To DS, From DS, More Fragments, Retry, Power Management, More Data, Protected Frame, and Order. See Figure 3–6.

Protocol Version subfield

The Protocol Version subfield is 2 bits long. It is used to identify the version of the IEEE 802.11 MAC protocol used to construct the frame. This version field is set to zero in the current version of the standard. All other values are reserved. The intended operation of this subfield is to allow a STA receiving a frame to determine whether the frame was constructed with a version of the protocol that the STA understands. If the protocol version indicates that the frame was constructed by a version of the IEEE 802.11 MAC protocol that the STA does not understand, the STA must discard the frame and not generate any response on the medium or any indication to higher layer protocols that the frame was received.

Type and Subtype subfields

The Type and Subtype fields identify the function of the frame and which other MAC header fields are present in the frame. There are three frame types: control, data, and management. The fourth frame type is currently reserved. Within each frame type, there are several subtypes. Table 3–1 provides the complete list of frame type and subtype combinations.

To DS and From DS subfields

The To DS subfield is 1 bit long. It is used only in data frames to indicate that the frame is destined for the DS. It will be set in every data frame sent from a mobile STA to the AP. This bit is zero in all other types of frames.

B0 B1	B2 B3	B4 B7	B8	B9	B10	B11	B11	B13	B14	B15
Protocol Version	Type	Subtype	To DS	From DS	More Frag	Retry	Pwr Mgt	More Data	Protected Frame	Order
2	2	4	1	1	1	1	1	1	1	1

Bits

Figure 3–6: Frame Control field

Table 3–1: Frame type and subtype

Type value b3 b2	Type description	Subtype value b7 b6 b5 b4	Subtype description
00	Management	0000	Association Request
00	Management	0001	Association Response
00	Management	0010	Reassociation Request
00	Management	0011	Reassociation Response
00	Management	0100	Probe Request
00	Management	0101	Probe Response
00	Management	0110–0111	Reserved
00	Management	1000	Beacon
00	Management	1001	Announcement traffic indication message (ATIM)
00	Management	1010	Disassociation
00	Management	1011	Authentication
00	Management	1100	Deauthentication
00	Management	1101	Action
00	Management	1110–1111	Reserved
01	Control	0000–0111	Reserved
01	Control	1000	BlockAckReq
01	Control	1001	BlockAck
01	Control	1010	Power Save (PS)-Poll
01	Control	1011	Request To Send (RTS)
01	Control	1100	Clear To Send (CTS)
01	Control	1101	Acknowledgment (ACK)

Table 3–1: Frame type and subtype *(Continued)*

Type value b3 b2	Type description	Subtype value b7 b6 b5 b4	Subtype description
01	Control	1110	Contention Free (CF)-End
01	Control	1111	CF-End + CF-ACK
10	Data	0000	Data
10	Data	0001	Data + CF-ACK
10	Data	0010	Data + CF-Poll
10	Data	0011	Data + CF-ACK + CF-Poll
10	Data	0100	Null Function (no data)
10	Data	0101	CF-ACK (no data)
10	Data	0110	CF-Poll (no data)
10	Data	0111	CF-ACK + CF-Poll (no data)
10	Data	1000	QoS Data
10	Data	1001	QoS Data + CF-ACK
10	Data	1010	QoS Data + CF-Poll
10	Data	1011	QoS Data + CF-ACK + CF-Poll
10	Data	1100	QoS Null (no data)
10	Data	1101	Reserved
10	Data	1110	QoS CF-Poll (no data)
10	Data	1111	QoS CF-Poll + CF-ACK (no data)
11	Reserved	0000–1111	Reserved

The From DS subfield is 1 bit long. It is also used only in data frames to indicate that the frame is being sent from the DS. This bit will be set in every data frame sent from the AP to a mobile STA. This bit is zero in all other types of frames.

There are four allowable combinations for these two subfields. When both subfields are zero, the frame is a direct communication between two mobile STAs. When the To DS subfield is one and the From DS subfield is zero, the frame is a transmission from a mobile STA to an AP. When the To DS subfield is zero and the From DS subfield is one, the frame is a transmission from an AP to a mobile STA. The final combination, when both subfields are one, is used for a special case where an IEEE 802.11 WLAN is being used as the DS. This last case is referred to as a *wireless DS*.

The reason for the special case of a wireless DS is to allow the DS to occupy the same medium as the BSS. If this case did not exist, there could be confusion about the addressing of the frames. When both subfields are one, the frame is being sent (distributed) from one AP to another, over the wireless medium.

More Fragments subfield

The More Fragments subfield is 1 bit long. This subfield is used to indicate that this frame is not the last fragment of a data or management frame that has been fragmented. This subfield is zero in the last fragment of a data or management frame that has been fragmented, in all control frames, and in any data or management frame that is not fragmented.

Retry subfield

The Retry subfield is 1 bit long. It is used to indicate if a data or management frame is being transmitted for the first time or if it is a retransmission. When this subfield is zero, the frame is being sent for the first time. When this subfield is one, the frame is a retransmission. To enable a receiving MAC to filter out duplicate received frames, the MAC uses this subfield along with the Sequence Number subfield of the Sequence Control field.

Power Management subfield

The Power Management subfield is 1 bit long. A mobile STA uses the Power Management subfield to announce its power management state. The value of the subfield indicates the power management state that the STA will enter when a successful frame exchange is completed. When this subfield is zero, the STA is in the active mode and will be available for future communication. When this subfield is one, the STA will be entering the power management mode and will not be available for future communication. This subfield must contain the same value for all frames transmitted by the STA during a single frame exchange. The STA may not change its power management state until it has completed a successful frame exchange. A successful frame exchange is the complete two-way or 4-Way Handshake, including the correct reception of an acknowledgment.

More Data subfield

The More Data subfield is 1 bit long. The AP uses this subfield to indicate to a mobile STA that there is at least one frame buffered at the AP for the mobile STA. When this subfield is one, there is at least one frame buffered at the AP for the mobile STA. When this subfield is zero, there are no frames buffered at the AP for the mobile STA. A mobile STA that is polled by the PC during a CFP also may use this subfield to indicate to the PC that there is at least one more frame buffered at the mobile STA to be sent to the PC. In multicast frames, the AP may also set this subfield to one to indicate that there are more multicast frames buffered at the AP.

Protected Frame subfield

The Protected Frame subfield is 1 bit long. When this subfield is one, the frame body of the MAC frame has been encrypted. This subfield may be set to one only in data frames and management frames of subtype Authentication. It is zero in all other frame types and subtypes.

Order subfield

The Order subfield is 1 bit long. When this subfield is one, the content of the data frame was provided to the MAC with a request for strictly ordered service. This subfield provides information to the AP and DS to allow this service to be delivered.

Duration/ID field

The Duration/ID field is 16 bits long. It alternately contains duration information for updating the NAV or a short ID, called the *association ID (AID)*, used by a mobile STA to retrieve frames that are buffered for it at the AP. Only the power-save poll (PS-Poll) frame contains the AID. In that frame, the AID is aligned in the 14 least significant bits (LSBs) of the field. The 2 most significant bits (MSBs) of the field are both set to one in the PS-Poll frame. Because of other limitations in the protocol, the maximum allowable value for the AID is 2007. All values larger than 2007 are reserved.

When bit 15 of the field is zero, the value in bits 14–0 represent the remaining duration of a frame exchange. This value is used to update the NAV and prevent a STA receiving this field from beginning a transmission that might cause corruption of the ongoing transmission.

The value of the Duration/ID field is set to 32, 768, i.e., bit 15 is one and all other bits are zero, in all frames transmitted during the CFP. This value is chosen to allow a STA that did not receive the beginning of the CFP to recognize that a CFP is ongoing and to set its NAV to a value large enough not to interfere with the CFP.

Other than the AID values, where bits 15 and 14 are set to one, all other values in this field are reserved.

Address fields

The MAC frame format contains four address fields. Any particular frame type may contain one, two, three, or four address fields. In IEEE 802.11, the address format is the familiar IEEE 48-bit address, normally used to identify the source and destination MAC addresses contained in a frame, as in

IEEE 802.3. In addition to the source address (SA) and destination address (DA), the IEEE 802.11 base standard defines three additional address types: the transmitter address (TA), the receiver address (RA), and the BSS identifier (BSSID). These additional address types are used in IEEE 802.11 to facilitate the level of indirection that allows transparent mobility and to provide a mechanism for filtering multicast frames. The position of the address in the address fields determines its function.

An IEEE 48-bit address comprises three fields: a 1-bit Individual/Group field, a 1-bit Universal/Local field, and a 46-bit Address field. The Individual/Group field defines whether the address is that of a single MAC or of a group of MACs. When the Individual/Group field is one, the remainder of the address identifies a group. If, in addition, all of the remaining bits in the address are set to one, the group is the broadcast group and includes all STAs. When the Individual/Group field is zero, the remainder of the address identifies a single MAC. The Universal/Local field defines whether the address is administered globally by the IEEE or locally. When the Universal/Local field is zero, the address is a globally administered address and should be unique. When the Universal/Local field is one, the address is locally administered and may not be unique.

BSS identifier (BSSID)

The BSSID is a unique identifier for a particular BSS of an IEEE 802.11 WLAN. Its format is identical to the format of an IEEE 48-bit address. In an infrastructure BSS, the BSSID is the MAC address of the AP. Using the MAC address of the AP for the BSSID ensures that the BSSID will be unique and also simplifies the address processing in the AP. In an IBSS, the BSSID is a locally administered, individual address that is generated randomly by the STA that starts the IBSS. The generation of this address from a random number provides some assurance that the address will be unique. However, there is a finite probability that the address generated is not unique. In both infrastructure BSSs and IBSSs, the BSSID must be an individual address. There is only one case where a group address is used: in a Probe Request frame. The use of the BSSID in the Probe Request frame is discussed in the description of that frame type.

Transmitter address (TA)

The TA is the address of the MAC that transmitted the frame onto the wireless medium. This address is always an individual address. The TA is used by STAs receiving a frame to identify the STA to which any responses in the MAC frame exchange protocol will be sent.

Receiver address (RA)

The RA is the address of the MAC to which the frame is sent over the wireless medium. This address may be either an individual or group address.

Source address (SA)

The SA is the address of the MAC that originated the frame. This address is always an individual address. This address does not always match the address in the TA field because of the indirection that is performed by the DS of an IEEE 802.11 WLAN. The SA field should be used to identify the source of a frame when indicating a frame has been received to higher layer protocols.

Destination address (DA)

The DA is the address of the final destination to which the frame is sent. This address may be either an individual or group address. This address does not always match the address in the RA field because of the indirection that is performed by the DS.

Sequence Control field

The Sequence Control field is a 16-bit field comprising two subfields: a 12-bit sequence number and a 4-bit fragment number. As a whole, this field is used to allow a receiving STA to eliminate duplicate received frames.

Sequence Number subfield

The Sequence Number subfield contains a 12-bit number assigned sequentially by the sending STA to each MSDU. This sequence number is incremented after each assignment and wraps back to zero when incremented from 4095. The sequence number for a particular MSDU is transmitted in every data frame associated with the MSDU. It is constant over all transmissions and retransmissions of the MSDU. If the MSDU is fragmented, the sequence number of the MSDU is sent with each frame containing a fragment of the MSDU.

Fragment Number subfield

The Fragment Number subfield contains a 4-bit number assigned to each fragment of an MSDU. The first, or only, fragment of an MSDU is assigned a fragment number of zero. Each successive fragment is assigned a sequentially incremented fragment number. The fragment number is constant in all transmissions or retransmissions of a particular fragment.

Frame Body field

The Frame Body field contains the information specific to the particular data or management frames. This field is variable length. It may be as long as 2304 bytes, without wired equivalent privacy (WEP) encryption, or 2312 bytes, when the frame body is encrypted using WEP. The maximum length when using 802.11i is described in Chapter 4. The value of 2304 bytes as the maximum length of this field was chosen to allow an application to send 2048-byte pieces of information, which can then be encapsulated by as many as 256 bytes of upper layer protocol headers and trailers.

FCS field

The FCS field is 32 bits long. It contains the result of applying the ITU CRC-32 polynomial to the MAC header and frame body. The CRC-32 polynomial is represented by the following equation:

$$G(x) = x^{32} + x^{26} + x^{23} + x^{22} + x^{16} + x^{12} + x^{11} + x^{10} + x^8 + x^7 + x^5 + x^4 + x^2 + x + 1$$

This polynomial is the same one used in other IEEE 802 LAN standards. The FCS in an IEEE 802.11 frame is generated in the same way as it is in IEEE 802.3.

CONTROL FRAME SUBTYPES

There are six control frame subtypes: Request To Send (RTS), Clear To Send (CTS), Acknowledge (ACK), Power Save Poll (PS-Poll), Contention-Free End (CF-End), and CF-End plus ACK (CF-End + ACK). A description of these frames follows.

Request to Send (RTS) [control] frame

The RTS frame is 20 bytes long. It comprises the Frame Control field, the Duration/ID field, two address fields, and the FCS field. The purpose of this frame is to transmit the duration information to STAs in the neighborhood of the transmitter in order that the STAs receiving the RTS frame will update their NAV to prevent transmissions from colliding with the data or management frame that is expected to follow. The RTS is also Frame 1 in a 4-Way Handshake between the transmitter and the receiver. See Figure 3–7.

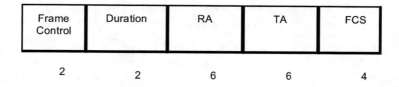

Frame Control	Duration	RA	TA	FCS
2	2	6	6	4

Figure 3–7:   RTS frame

The RA field identifies the individual MAC that is the immediate intended recipient of the frame. In an RTS frame, the RA is always an individual address. The TA field identifies the source of the transmission. It is used by the STA addressed by the RA to form the CTS frame that is the response to the RTS frame. The duration information conveyed by this frame is a measure of the amount of time required to complete the four-way frame exchange. The value of the duration is the length of time to transmit a CTS frame, the data or management frame, the ACK frame, and the three SIFS intervals between the RTS frame and the CTS frame, between the CTS frame and the data or management frame, and between the data or management frame and the ACK frame. The duration is measured in microseconds. Fractional microseconds are always rounded up to the next larger integer value.

Clear to Send (CTS) [control] frame

The CTS frame is 14 bytes long. It comprises the Frame Control field, the Duration/ID field, one address field, and the FCS field. The purpose of this frame is to transmit the duration information to STAs in the neighborhood of the STA intended to receive the expected data or management frame in order that the STAs receiving the CTS frame will update their NAVs to prevent transmissions from colliding with the data or management frame that is expected to follow. The value of the duration filed in the CTS frame is determined by subtracting one SIFS interval and the transmission time of the CTS frame from the value of the duration field received in the RTS frame. The CTS is Frame 2 in a 4-Way Handshake between the transmitter and the receiver. See Figure 3–8.

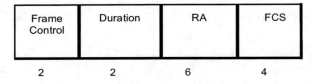

Frame Control	Duration	RA	FCS
2	2	6	4

Figure 3–8:   CTS frame

The RA field identifies the individual MAC address of the STA to which the CTS frame is sent. In the CTS frame, the RA is always an individual address. The RA value is taken directly from the TA field of the preceding RTS frame. The duration information conveyed by this frame is a measure of the time required to complete the 4-Way Handshake. The value of the duration is the length of time to send the subsequent data or management frame, the ACK frame, and one SIFS interval. The duration value is calculated by subtracting the length of time to transmit a CTS frame and one SIFS interval from the duration that was received in the RTS frame. The duration is measured in microseconds. Fractional microseconds are always rounded up to the next larger integer value.

Acknowledge (ACK) [control] frame

The ACK frame is 14 bytes long. It comprises the Frame Control field, the Duration/ID field, one address field, and the FCS field. The purpose of this frame is two-fold. First, the ACK frame transmits an acknowledgment to the sender of the immediately previous data, management, or PS-Poll frame that the frame was received correctly. This acknowledgment informs the sender of the frame of the frame's receipt and eliminates the requirement for retransmission by the sender. Second, the ACK frame is used to transmit the duration information for a fragment burst to STAs in the neighborhood of the STA intended to receive the fragments. In this case, it performs exactly as the CTS frame. The ACK frame is Frame 4 in the 4-Way Handshake between transmitter and receiver. See Figure 3–9.

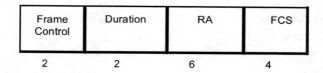

Frame Control	Duration	RA	FCS
2	2	6	4

Figure 3–9: ACK frame

The RA field identifies the individual MAC address of the STA to which the ACK frame is sent. In the ACK frame, the RA is always an individual address. The RA value is taken directly from the Address 2 field of the immediately preceding data, management, or PS-Poll frame.

The value of the duration information is zero if the ACK frame is an acknowledgment of a PS-Poll frame or is an acknowledgment of a management or data frame where the More Fragments subfield of the Frame Control field is zero. The value of the duration information is the time to transmit the subsequent data or management frame, an ACK frame, and two SIFS intervals if the acknowledgment is of a data or management frame where the More Fragments subfield of the Frame Control field is one. In this case, the duration may be calculated by subtracting the length of time to transmit the ACK frame and one SIFS interval from the duration value received in the immediately preceding data or management frame. The duration value is measured in microseconds. Fractional microseconds are always rounded up to the next integer value.

Power Save Poll (PS-Poll) [control] frame

The PS-Poll frame is 20 bytes long. It comprises the Frame Control field, the Duration/ID field, two address fields, and the FCS field. The purpose of this frame is to request that an AP deliver a frame that has been buffered for a mobile STA while it was in a power-saving mode.

The BSSID identifies the AP to which this frame is directed. This BSSID should be the same BSSID as that to which the sending STA has previously associated. The BSSID in the PS-Poll frame is always an individual address. The TA is the MAC address of the mobile STA that is sending the PS-Poll frame. The duration/ID value is the AID value that was given to the mobile STA upon association with the BSS. Even though this frame does not include any explicit duration information, every mobile STA receiving a PS-Poll frame will update its NAV with a value that is the length of time to transmit an ACK frame and a single SIFS interval. This action allows the ACK frame that follows the PS-Poll frame to be sent by the AP with a very small probability that it will collide with frames from mobile STAs.

For the AP response to this frame, see Chapter 9.

Contention-Free End (CF-End) and CF-End plus ACK (CF-End + ACK) [control] frames

The CF-End and CF-End + ACK frames are 20 bytes long. Each frame comprises the Frame Control field, the Duration/ID field, two address fields, and the FCS field. The purpose of these frames is to conclude a CFP and to release STAs from the restriction imposed during a CFP to prevent competition for access to the medium. Additionally, the CF-End + ACK frame is used to acknowledge the last transmission received by the PC. This frame is sent by the PC as the last frame in the CFP.

The RA is the broadcast group address, as this frame is intended to be received by every STA in the BSS. The BSSID is the MAC address of the AP where the PC resides. The duration value is zero to ensure that the NAVs of all STAs receiving this frame will be reset to zero.

DATA FRAME SUBTYPES

There are eight data frame subtypes in two groups. The first group is simple data, Data with Contention-Free Acknowledgment (CF-ACK), Data with Contention-Free Poll (CF-Poll), and Data with CF-ACK and CF-Poll frames. The second group is Null Function, CF-ACK, CF-Poll, and CF-ACK + CF-Poll frames. The first group of data frames actually carry a nonzero number of data bytes. The second group of data frames carry no data bytes at all.

The data frame is variable in length and carries the MSDU requested to be delivered by the upper layer protocols. The minimum length of the data frame is 29 bytes. The maximum length of the data frame is 2346 bytes. It comprises the Frame Control field, the Duration/ID field, up to four address fields, the Sequence Control field, the Frame Body field, and the FCS field. See Figure 3–10.

Simple data frame

The simple data frame encapsulates the upper layer protocol packets, delivering them from one IEEE 802.11 STA to another. It may appear in both the contention period and contention-free period (CFP).

The Duration/ID field contains a value that is measured in microseconds from the end of the data frame and is sufficient to protect the transmission of a subsequent ACK frame if the More Fragments subfield of the Frame Control field is zero. This value is determined by adding the length of a SIFS interval to the transmission time of the expected ACK frame. The transmission time of the ACK frame will make this value variable, dependent only on the data rate of the ACK frame. The data rate at which the ACK frame is transmitted is determined by the multirate rules of IEEE 802.11, which state that the ACK frame is sent at the same rate as the frame it is acknowledging, unless that rate is not in the basic rate set for the BSS. If the data rate of the frame that the ACK frame is acknowledging is not in the BSS basic rate set, the ACK frame will be sent at the highest rate in the BSS basic rate set that is less than the rate of the frame it is acknowledging. The use of any value other than the value determined from the multirate rules will lead to failure of DCF fairness of access to the medium.

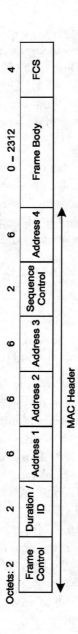

Figure 3–10: Data frame

If the More Fragments subfield of the Fame Control field is one, the value of the Duration/ID field is determined as if the frame is an RTS frame. If the data frame is sent to a multicast address, the duration/ID value is zero, as no ACK frame is expected in response.

The use of the four address fields in the data frame are dependent on two things: whether the BSS to which the transmitting STA belongs is an IBSS or an infrastructure BSS and whether the transmitting and/or receiving STAs are part of the DS. Table 3–2 identifies the functions of each of the address fields for the four possible cases of the To DS and From DS bits.

Table 3–2: Functions of address fields in data frames

Function	To DS	From DS	Address 1	Address 2	Address 3	Address 4
IBSS	0	0	RA = DA	TA = SA	BSSID	N/A
From the AP	0	1	RA = DA	TA = BSSID	SA	N/A
To the AP	1	0	RA = BSSID	TA = SA	DA	N/A
Wireless DS	1	1	RA	TA	DA	SA

The Address 1 field is always used to perform RA matching decisions. If the address in the Address 1 field is an individual address, the STA will receive the frame and indicate it to higher layer protocols only if the address matches its own. When the Address 1 field contains a group address, the BSSID is also checked to determine that the frame was sent from a STA in the same BSS to which the receiving STA belongs.

The Address 2 field is used to identify the sender of the frame. The content of this field is used to direct any required acknowledgment back to the sender. This address is always an individual address.

The Address 3 field carries additional information for frame filtering or forwarding by the DS. A frame received by a mobile STA from an AP will use the address in this field to indicate the SA of the frame to higher layer protocols. A frame received by an AP from a mobile STA will use the address in this field as the DA of the frame for DS forwarding decisions. In the case of the wireless DS, this field contains the DA of the frame that was originally received by the AP.

The Address 4 field is used only in a wireless DS as one AP forwards a frame to another AP. This field is not present in any other data frame. In this case, the SA from the original frame received by the AP is contained in this field. The address in this field will be placed by the DS into the Address 3 field of a frame that is delivered by the AP to a mobile STA or into the SA field of a frame that is passed onto a wired network.

The DA is the destination of the MSDU in the Frame Body field. The SA is the address of the MAC entity that initiated the MSDU in the Frame Body field. The RA is the address of the STA contained in the AP in the wireless DS that is the next immediate intended recipient of the frame. The TA is the address of the STA contained in the AP in the wireless DS that is transmitting the frame. The BSSID of the data frame is determined as follows:

a) If the STA is an AP or is associated with an AP, the BSSID is the address currently in use by the STA contained in the AP.

b) If the STA is a member of an IBSS, the BSSID is the BSSID of the IBSS.

Data with Contention-Free Acknowledgment (Data + CF-ACK) frame

The Data + CF-ACK frame is identical to the simple data frame, with the following exceptions. The Data + CF-ACK frame may be sent only during a CFP. It is never used in an IBSS. The acknowledgment carried in this frame is acknowledging the previously received data frame, which may not be associated with the address of the destination of the current frame.

Data with Contention-Free Poll (Data + CF-Poll) frame

The Data + CF-Poll frame is identical to the simple data frame, with the following exceptions. The Data + CF-Poll frame may be sent only by the PC during a CFP. This frame is never sent by a mobile STA. It is never used in an IBSS. This frame is used by the PC to deliver data to a mobile STA and simultaneously request that the mobile STA send any buffered data frame when the current reception is completed.

Data + CF-ACK + CF-Poll frame

The Data + CF-ACK + CF-Poll frame is identical to the simple data frame, with the following exceptions. The Data + CF-ACK + CF-Poll frame may be sent only by the PC during a CFP. This frame is never sent by a mobile STA. It is never used in an IBSS. This frame combines the functions of both the Data + CF-ACK and Data + CF-Poll frames into a single frame.

Null Function (no data) frame

The Null Function frame, a data frame that contains no frame body and thus no data, is used to allow a STA that has nothing to transmit to be able to complete the frame exchange necessary for changing its power management state. The sole purpose for this frame is to carry the power management bit in the Frame Control field to the AP when a STA changes to a low power operating state. More detail on power management operation is provided in Chapter 9.

Contention-Free Acknowledgment (CF-ACK) (no data) frame

The CF-ACK frame may be used by a mobile STA that has received a data frame from the PC during the CFP to acknowledge the correct receipt of that frame. Because this frame uses the Null function frame format and is 28 bytes long, it is more efficient to use a shorter acknowledgment control frame (i.e., ACK frame). Either frame will provide the required acknowledgment to the PC.

Contention-Free Poll (CF-Poll) (no data) frame

The CF-Poll frame is used by the PC to request that a mobile STA send a pending data frame during the CFP. The PC will send this frame, rather than the Data + CF-Poll, when it has no data to be sent to the mobile STA. This frame uses the Null Function frame format.

CF-ACK + CF-Poll (no data) frame

The CF-ACK + CF-Poll is used by the PC to acknowledge a correctly received frame and to solicit a pending frame from a mobile STA. The acknowledgment and the solicitation may be intended for disparate mobile STAs. This frame uses the Null Function frame format.

MANAGEMENT FRAME SUBTYPES

IEEE 802.11 is different from many of the other IEEE 802 standards because it includes very extensive management capabilities defined at the MAC level. One of the four MAC frame types is dedicated to management frames. There are 11 distinct management frame subtypes [Beacon, Probe Request, Probe Response, Authentication, Deauthentication, Association Request, Association Response, Reassociation Request, Reassociation Response, Disassociation, and Announcement Traffic Indication Message (ATIM)] in addition to the general Action management frame. All management frames include the Frame Control field, the Duration field, three address fields, the Sequence Control field, the Frame Body field, and the FCS field.

The frame body of a management frame carries information both in fixed-length mandatory components defined as *fixed fields* and in usually variable-length mandatory and optional components defined as *information elements* that are dependent on subtype. The information element is a flexible data structure that contains an information element ID, a length, and the content of the information element. Information elements occur in the frame body in order of increasing IDs. This arrangement and the data structure itself allow for the flexible extension of the management frames to include new functionality without affecting older implementations. This flexibility can be accomplished because older implementations will be able to understand the older elements and will ignore elements with new IDs. Because the length of the element is part of the data structure, an older implementation can skip over newer elements without needing to understand the content of the element. See Figure 3–11.

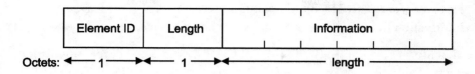

Figure 3–11: Information element

All implementations should be prepared to find information elements in any management frame, even if those frames do not currently carry any information elements found in the IEEE 802.11 standard and its amendments. Even though the standard currently provides an order of inclusion of information elements in each frame, this order may not persist as future amendments are made to the standard. Prudent implementers will parse all management frames in a way that will correctly interpret information elements encountered in any order in any management frame. Dependence on a particular order of information elements, or the presence of only a fixed set of information elements, in a frame is a very poor design and should be avoided at all cost. This statement is particularly true with the definition of the vendor-specific information element, for which the standard does not provide any guidance as to its presence or order in any frame.

Descriptions of the fixed fields and information elements found in the frame body follow the descriptions of the types of management frames.

Beacon [management] frame

The Beacon frame is transmitted periodically to allow mobile STAs to locate and identify a BSS. The information in a Beacon frame allows a mobile STA to locate the BSS (in time and PHY parameters) at any time in the future. The Beacon frame also conveys information to mobile STAs about frames that may be buffered during times of low-power operation. The Beacon frame includes the following fixed fields: Timestamp, Beacon Interval, and Capability Information. The Timestamp field is 64 bits long and contains the value of the STA's synchronization timer at the time that the frame was transmitted. The Beacon Interval field is the period of beacon transmissions. The period is measured in time units (TUs) of 1024 μs and known as the *beacon period.* The Beacon Interval field is 16 bits long. The Capability Information field is 16 bits long and identifies the capabilities of the STA.

The information elements in a Beacon frame are Service Set Identity (SSID), Supported Rates, one or more PHY Parameter Sets, an optional CF Parameter Set, an optional IBSS Parameter Set, and an optional Traffic Indication Map (TIM). For a STA implementing IEEE 802.11d, the Beacon frame will include a Country information element and may include Hopping Pattern Parameters and Hopping Pattern Table information elements. For a STA implementing IEEE P802.11e, the Beacon frame will include QBSS Load and EDCA Parameter Set information elements. For a STA implementing IEEE 802.11g, the Beacon frame may contain ERP Information information element. For a STA implementing IEEE 802.11h, the Beacon frame may include Power Constraint, Supported Channels, Channel Switch Announcement, and Quiet information elements. For a STA implementing IEEE 802.11i, the Beacon frame may include the RSN information element.

Probe Request and Probe Response [management] frames

The Probe Request frame is transmitted by a mobile STA attempting to quickly locate an IEEE 802.11 WLAN. It may be used to locate a WLAN with a particular SSID or to locate any WLAN. The Probe Request frame contains two information elements: SSID and Supported Rates. When the SSID has a length greater than zero, the Probe Request frame is a *directed* request, i.e., it is asking for any device that beacons with the particular SSID in the Probe Request frame to respond with a Probe Response frame. When the SSID length is zero, the Probe Request frame is a *broadcast* (or *wild card*) request, i.e., it is asking for all devices sending beacons to respond with a Probe Response frame. Generally, the BSSID in the Probe Request frame is the *broadcast* BSSID. This type of frame requests that all STAs that match the SSID field respond with a Probe Response frame. The BSSID field can also be an individual BSSID. When the BSSID field contains an individual BSSID value, the Probe Request frame is seeking responses only from that individual BSSID. The RA (in the Address 1 field) of the Probe Request frame is always the broadcast address.

The effect of receiving a Probe Request frame is to cause the STA to respond with a Probe Response frame if the receiver was the last STA in the BSS to transmit a Beacon frame. In an infrastructure BSS, the AP will always respond to the Probe Request frames. In an IBSS, the mobile STA that sent the latest Beacon frame will respond. In an IBSS, this process may result in more than one STA responding to the Probe Request frame because the beaconing process in an IBSS may result in more than one STA sending a Beacon frame after each target beacon transmission time (TBTT).

The Probe Response frame contains nearly all the same information as a Beacon frame. The Probe Response frame includes the following fixed fields: Timestamp, Beacon Interval, and Capability Information. It also includes the SSID, Supported Rates, one or more PHY Parameter Sets, the optional CF Parameter Set, and the optional IBSS Parameter Set. For a STA implementing IEEE 802.11d, the Probe Response frame will include a Country information element and may include any other information element defined by IEEE 802.11 and its amendments if those information elements were requested in a Request information element in the Probe Request frame. For a STA implementing IEEE P802.11e, the Probe Request frame will include QBSS Load and EDCA Parameters information elements. For a STA implementing IEEE 802.11g, the Probe Response frame may contain ERP Information information element. For a STA implementing IEEE 802.11h, the Probe Response frame may include Power Constraint, Supported Channels, Channel Switch Announcement, and Quiet information elements. For a STA implementing IEEE 802.11i, the Probe Response frame may also include the RSN information element.

Authentication [management] frame

The Authentication frame is used to conduct a multiframe exchange between STAs that ultimately results in the verification of the identity of each STA to the other, within certain constraints. The Authentication frame includes three fixed fields: Authentication Algorithm Number, Authentication Transaction Sequence Number, and Status Code. The Authentication frame may also include information elements. The presence of the status code and challenge text or other information elements in the Authentication frame is dependent on the algorithm used and the transaction sequence number.

Deauthentication [management] frame

The Deauthentication frame is used by a STA to notify another STA of the termination of an authentication relationship. The Deauthentication frame includes only a single fixed field: Reason Code.

Association Request and Association Response [management] frames

The Association Request and Association Response frames are used by a mobile STA to request an association with a BSS and for the success or failure of that request to be returned to the mobile STA. The Association Request frame includes two fixed fields: Capability Information and Listen Interval. There are also two information elements in the Association Request frame: SSID and Supported Rates.

The Association Response frame includes three fixed fields: Capability Information, Status Code, and AID. There is one information element in the Association Response frame: Supported Rates. It is important for a STA to examine the status code of failed associations and to respond in a reasonable fashion. The IEEE 802.11 standard does not specify how the status codes are to be used. However, in order to avoid problems such as continuously attempting to associate with an AP that is not accepting additional associations or attempting to associate with an AP that is "leaving," i.e., going out of service, a STA must recognize the reason for a failed association and modify its default association behavior accordingly.

Reassociation Request and Reassociation Response [management] frames

The Reassociation Request and Reassociation Response frames are used by a mobile STA that has been associated with a BSS and is now associating with another BSS with the same SSID. The Reassociation Request frame includes the same information as an Association Request frame, with the addition of a Current AP Address field.

The Reassociation Response frame is identical to the Association Response frame.

Disassociation [management] frame

The Disassociation frame is used by a STA to notify another STA of the termination of an association relationship. The Disassociation frame includes only a single fixed field: Reason Code. The reason must be examined when a STA receives a Disassociation frame. The IEEE 802.11 standard does not specify how the reason codes are to be interpreted. However, a well-behaved STA will include the reason for a disassociation into its decision-making process when choosing a new AP.

Announcement Traffic Indication Message (ATIM) [management] frame

The ATIM frame is used by mobile STAs in an IBSS to notify other mobile STAs in the IBSS that may have been operating in low power modes that the sender of the ATIM frame has traffic buffered and waiting to be delivered to the STA addressed in the ATIM frame. The ATIM frame does not include any fixed fields or information elements.

Action [management] frame

The Action frame is a general management "envelope," used to carry other management commands, requests, and responses. These uses are described further in Chapter 5 and Chapter 6.

COMPONENTS OF THE MANAGEMENT FRAME BODY

The components of the management frame body comprise fixed fields and information elements.

Fixed fields

There are 10 fixed fields that may be used in the frame body of management frames: AID, Authentication Algorithm Number, Authentication Transaction Sequence Number, Beacon Interval, Capability Information, Current AP Address, Listen Interval, Reason Code, Status Code, and Timestamp.

AID field

The AID field is 16 bits long. It contains an arbitrary number assigned by the AP when a STA associates with a BSS. The format of this field is identical to that of the Duration/ID field of the MAC frame header. The values allowed for this field are 1 through 2007 in the 14 LSBs. The 2 MSBs must both be set to 1. The numeric value in the 14 LSBs of this field are used by the mobile STA to identify which bit in a TIM information element indicates that the AP has frames buffered for the mobile STA.

Authentication Algorithm Number field

The Authentication Algorithm Number field is 16 bits long. It contains a number identifying the authentication algorithm to be used to complete an authentication transaction. Currently only two values are defined for this field; all other values are reserved for future standardization. When the authentication algorithm number is zero, the algorithm to be used is open system authentication. When the authentication algorithm number is one, the algorithm to be used is shared key authentication. Each of the algorithms is described in Chapter 9.

Authentication Transaction Sequence Number field

The Authentication Transaction Sequence Number field is 16 bit s long. It is used to track the progress of an authentication transaction. The authentication sequence transaction number is increased sequentially with each Authentication frame exchanged during the transaction. The initial value for the authentication transaction sequence number is 1. The authentication transaction sequence number may not take the value 0.

Beacon Interval field

The Beacon Interval field is 16 bits long. It is a numeric value indicating the typical amount of time that elapses between beacon frame transmissions. The time interval is measured in TU. One TU is 1024 µs.

Capability Information field

The Capability Information field is 16 bits long. It indicates the capabilities of a STA. See Figure 3–12. The original subfields of the Capability Information field are ESS, IBSS, CF Pollable, CF-Poll Request, and Privacy. In IEEE 802.11b, three subfields are added: Short Preamble, PBCC, and Channel Agility. In IEEE 802.11h, one subfield is added: Spectrum Management. In IEEE P802.11e, more subfields are added: QoS, APSD, Delayed Block ACK, and Immediate Block ACK. In IEEE 802.11g, two subfield are added: Short Slot Time and DSSS-OFDM. One bit in the Capability Information field remains reserved.

The rules for using the Capability Information field are complex and depend on whether the STA is an AP or mobile STA or whether it is part of an IBSS. Some subfields are significant only in certain management frames.

The ESS and IBSS subfields are significant only in Beacon, Probe Response, Association Request, Reassociation Request, Association Response, and Reassociation Response frames. An AP always sets the ESS subfield to one and the IBSS subfield to zero in Beacon, Probe Response, Association Response, and Reassociation Response frames. A mobile STA in an IBSS always sets the ESS subfield to zero and the IBSS subfield to one. When requesting association or reassociation, a mobile STA sets the ESS and IBSS bits to match the type of WLAN with which it is attempting to associate or reassociate, i.e., to associate with an ESS, the ESS bit must be set to one and the IBSS bit must be set to zero.

The CF Pollable and CF-Poll Request subfields are significant in Beacon, Probe Response, Association Request, Association Response, Reassociation Request, and Reassociation Response frames. A mobile STA will set these subfields in Association Request and Reassociation Request frames to indicate its contention-free capability and to request that it be placed on the polling list of the PC. Table 3–3 is taken from the IEEE 802.11 standard to describe the functions of these subfields in a mobile STA.

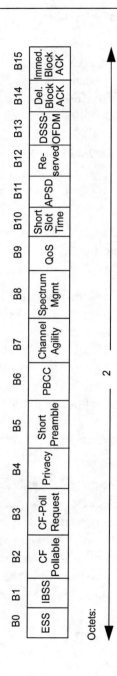

Figure 3–12: Capability Information field

Table 3–3:   Functions of subfields in a mobile STA

CF Pollable	CF Poll Request	Meaning
0	0	STA is not CF-pollable
0	1	STA is CF-pollable, not requesting to be placed on the CF-polling list
1	0	STA is CF-pollable, requesting to be placed on the CF-polling list
1	1	STA is CF-pollable, requesting never to be polled

An AP will set the CF Pollable and CF-Poll Request subfields in Beacon, Probe Response, Association Response and Reassociation Response frames to indicate the capability of the PC, if any. Table 3–4 is taken from the IEEE 802.11 standard to describe the functions of the subfields in an AP.

Table 3–4:   Functions of subfields in an AP

CF Pollable	CF Poll Request	Meaning
0	0	No PC at AP
0	1	PC at AP for delivery only (no polling)
1	0	PC at AP for delivery and polling
1	1	Reserved

In both of these tables, it can be seen that the names of the subfields do not quite match with the functions that are described in the tables. Originally, these two subfields were independent. However, during the development process of the standard, a case was made for including the ability for a STA that does have contention-free polling capabilities to request that it never be polled, but be treated as if it did not have contention-free capability. This left three cases of a STA with contention-free capabilities, but only two rows of the table to logically indicate them. At this point the interpretation of the two subfields was changed to be a 2-bit label. It was also decided that the interpretation of the table for the capabilities of the AP be changed similarly.

The Privacy subfield is transmitted by the AP in Beacon, Probe Response, Association Response, and Reassociation Response frames. This subfield indicates that the use of confidentiality [either wired equivalent privacy (WEP) or robust security network (RSN)] is required for all data type frames, when set to one. When set to zero, this subfield indicates that the use of confidentiality is not required. When IEEE 802.11i robust security is used, the Privacy subfield is always sent as zero by STAs and ignored by APs. See Chapter 4 for additional information on IEEE 802.11i RSN operation.

The Short Preamble subfield is transmitted by an AP or a mobile STA in an IBSS in Beacon, Probe Response, Association Response, and Reassociation Response frames to indicate the availability of the short preamble option when using an IEEE 802.11b PHY. When set to one, this subfield indicates that the use of short preambles is allowed in the BSS. When set to zero, this subfield indicates that the use of short preamble is not allowed in the BSS.

In a mobile STA that is not part of an IBSS, the Short Preamble subfield in Association Request and Reassociation Request frames indicates the capability of the STA to send and receive the short preambles of IEEE 802.11b.

The PBCC subfield is transmitted by an AP or a mobile STA in an IBSS in Beacon, Probe Response, Association Response, and Reassociation Response frames to indicate the availability of the packet binary convolutional code (PBCC) option when using an IEEE 802.11b PHY. When set to one, this subfield indicates that the use of PBCC is allowed in the BSS. When set to zero, this subfield indicates that the use of PBCC is not allowed in the BSS.

In a mobile STA that is not part of an IBSS, the PBCC subfield in Association Request and Reassociation Request frames indicates the capability of the STA to send and receive the PBCC of IEEE 802.11b.

The Channel Agility subfield indicates that the STA is using the channel agility option of IEEE 802.11b.

The Spectrum Management subfield indicates that the STA requires the use of IEEE 802.11h spectrum management.

The QoS subfield indicates that the STA supports IEEE P802.11e quality of service (QoS) extensions. Information elements in Beacon, Probe Response, Association Response, and Reassociation Response frames indicate the IEEE P802.11e functionality is supported and required. See Chapter 5 for more detail on IEEE P802.11e QoS operation.

The Short Slot Time subfield in Beacon, Probe Response, Association Response, and Reassociation Response frames sent by the AP indicates the length of the slot time that is currently in use in the BSS. STAs indicate their capability to support short slot times in the Association Request and Reassociation Request frames. When an association from a STA that does not support short slot times is accepted by an AP, the AP indicates that short slot times may not be used in the BSS by transmitting the value of the short slot time subfield as a zero at the next Beacon frame after accepting the association. All STAs associated with the AP must abide by the current setting of the Short Slot Time subfield as it is transmitted by the AP.

The APSD subfield indicates the ability of the AP to support the power management mode of operation called *automatic power save delivery*. This subfield is always transmitted as a zero by a STA. See Chapter 5 for additional information on QoS operation.

The DSSS-OFDM subfield indicates that IEEE 802.11g operation is permitted in the BSS. See Chapter 14 for additional information on IEEE 802.11g operation.

The Delayed Block ACK subfield indicates that the STA or AP is capable of supporting delayed block acknowledgment for QoS data frames. See Chapter 5 for additional information on block acknowledgment.

Current AP Address field

The Current AP Address field is 6 bytes long and is used to hold the address of the AP with which a mobile STA is currently associated when that mobile STA is attempting to reassociate. If the reassociation is successful, the AP with which the mobile STA has reassociated may use the current AP address to contact that AP.

Listen Interval field

The Listen Interval field is 16 bits long. The listen interval is used by a mobile STA to indicate to an AP how long the mobile STA may be in low power operating modes and unable to receive frames. The value in the Listen Interval field is in units of the beacon period. For example, a STA that wakes only on every tenth beacon would set this field to 10. The AP may use this field to determine the resources required to support the mobile STA and may refuse an association based on that information. When an AP allows a STA to associate, the AP commits to being able to buffer any frames that arrive for the STA for the length of time indicated by the listen interval.

Reason Code field

The Reason Code field is 16 bits long and indicates the reason for an unsolicited notification of disassociation or deauthentication. Table 3–5 is taken from the IEEE 802.11 standard to describe the allowable values for the reason code. An implementation must consider the value of the reason code in any frame in which it appears. Algorithms that fail to take the reason code into account are very likely to perform exactly the same action that caused the Disassociation or Deauthentication frame to be delivered with the reason code in the first place, receiving the same reason code again and again, ad infinitum. Acting on the reason code or providing information to the user that allows action to be taken to correct the underlying cause for receiving the reason code is a much more robust implementation.

Table 3–5:   Reason codes

Reason Code	Meaning
0	Reserved
1	Unspecified reason
2	Previous authentication no longer valid
3	Deauthenticated because sending STA is leaving (has left) IBSS or ESS

Table 3–5: Reason codes *(Continued)*

Reason Code	Meaning
4	Disassociated due to inactivity
5	Disassociated because AP is unable to handle all currently associated STAs
6	Class 2 frame received from nonauthenticated STA
7	Class 3 frame received from nonassociated STA
8	Disassociated because sending STA is leaving (has left) BSS
9	STA requesting (re)association is not authenticated with responding STA
10	Disassociated because the information in the Power Capability information element is unacceptable
11	Disassociated because the information in the Supported Channels information element is unacceptable
12	Reserved
13	Invalid information element
14	Message integrity field (MIC) failure
15	4-Way Handshake timeout
16	Group key update timeout
17	Information element in 4-Way Handshake different from (Re)Association Request/Probe Response/Beacon frame
18	Group cipher is not valid
19	Pairwise cipher is not valid
20	AKMP is not valid
21	Unsupported RSN information element version
22	Invalid RSN information element capabilities

Table 3–5:   Reason codes *(Continued)*

Reason Code	Meaning
23	IEEE 802.1X™ authentication failed
24	Cipher suite is rejected per security policy
25–31	Reserved
32	Disassociated for unspecified, QoS-related reason
33	Disassociated because QoS AP (QAP) lacks sufficient bandwidth for this QoS STA (QSTA)
34	Disassociated because of excessive number of frames that need to be acknowledged, but are not acknowledged for AP transmissions and/or poor channel conditions
35	Disassociated because QSTA is transmitting outside the limits of its TXOPs
36	Requested from peer QSTA as the QSTA is leaving the QoS BSS (QBSS) (or resetting)
37	Requested from peer QSTA as it does not want to use the mechanism
38	Requested from peer QSTA as the QSTA received frames using the mechanism for which a set up is required
39	Requested from peer QSTA due to time out
40–44	Reserved
45	Peer QSTA does not support the requested cipher suite

Status Code field

The Status Code field is 16 bits long and indicates the success or failure of a requested operation. The value zero indicates a successful operation. Any nonzero value indicates the requested operation failed for the indicated reason. Table 3–6 is taken from the IEEE 802.11 standard to describe the meaning of the status codes. An implementation must consider the value of the status code in any frame in which it appears. Algorithms that fail to take the status code into account are very likely to perform exactly the same action that caused the frame to be delivered with the status code in the first place, receiving the same status code again and again, ad infinitum. Acting on the status code or providing information to the user that allows action to be taken to correct the underlying cause for receiving the status code is a much more robust implementation.

Table 3–6:   Status codes

Status Code	Meaning
0	Successful
1	Unspecified failure
2–9	Reserved
10	Cannot support all requested capabilities in the Capability Information field
11	Reassociation denied due to inability to confirm that association exists
12	Association denied due to reason outside the scope of this standard
13	Responding STA does not support the specified authentication algorithm
14	Received an Authentication frame with authentication transaction sequence number out of expected sequence
15	Authentication rejected because of challenge failure
16	Authentication rejected due to timeout waiting for next frame in sequence
17	Association denied because AP is unable to handle additional associated STAs

Table 3–6: Status codes *(Continued)*

Status Code	Meaning
18	Association denied because requesting STA does not support all of the data rates in the BSSBasicRateSet parameter
19	Association denied because requesting STA does not support the short preamble option
20	Association denied because requesting STA does not support the PBCC modulation option
21	Association denied because requesting STA does not support the channel agility option
22	Association request rejected because spectrum management capability is required
23	Association request rejected because the information in the Power Capability information element is unacceptable
24	Association request rejected because the information in the Supported Channels information element is unacceptable
25	Association denied because requesting STA does not support the short slot time option
26	Association denied because requesting STA does not support the DSSS-OFDM option
27–31	Reserved
32	Unspecified QoS-related failure
33	Association denied because QAP has insufficient bandwidth to handle another QSTA
34	Association denied due to excessive frame loss rates and/or poor conditions on current operating channel
35	Association (with QBSS) denied because requesting STA does not support the QoS facility

Table 3–6:   Status codes (Continued)

Status Code	Meaning
36	Reserved
37	Request declined
38	Request not successful because one or more parameters have invalid values.
39	TS not created because the request cannot be honored. However, a suggested TSPEC is provided so that the initiating QSTA may attempt to set another TS With the suggested changes to the TSPEC.
40	Invalid information element
41	Group cipher not valid
42	Pairwise cipher not valid
43	AKMP not valid
44	Unsupported RSN information element version
45	Invalid RSN information element capabilities
46	Cipher suite rejected per security policy
47	TS not created. However, the HC may be capable of creating a TS, in response to a request, after the time indicated in the TS Delay information element.
48	Direct link not allowed in the BSS by policy
49	Destination STA not present within this QBSS
50	Destination STA not a QSTA
51–65 535	Reserved

Timestamp field

The Timestamp field is 64 bit long and is the value of the STA's TSFTIMER at the time a frame was transmitted. The Timestamp field is used in Beacon and Probe Response frames.

Information elements

Table 3–7 is a listing of the individual information elements defined by the IEEE 802.11 standard (including IEEE P802.11e) and their associated element IDs.

Table 3–7:   Information elements and associated element IDs

Element ID	Information Element
0	SSID
1	Supported Rates
2	FH Parameter Set
3	DS Parameter Set
4	CF Parameter Set
5	TIM
6	IBSS Parameter Set
7	Country
8	Hopping Pattern Parameters
9	Hopping Pattern Table
10	Request
11	QBSS Load
12	EDCA Parameter Set
13	Traffic Specification (TSPEC)

**Table 3–7:   Information elements and
associated element IDs *(Continued)***

Element ID	Information Element
14	Traffic Classification
15	Schedule
16	Challenge Text
17–31	Reserved for challenge text extension
32	Power Constraint
33	Power Capability
34	TPC Request
35	TPC Report
36	Supported Channels
37	Channel Switch Announcement
38	Measurement Request
39	Measurement Report
40	Quiet
41	IBSS DFS
42	ERP Information
43	TS Delay
44	TCLAS Processing
45	Reserved
46	QoS Capability
47	Reserved
48	RSN
49	Reserved

**Table 3–7:   Information elements and
associated element IDs *(Continued)***

Element ID	Information Element
50	Extended Supported Rates
51–220	Reserved
221	Vendor Specific
222–255	Reserved

SSID information element

The SSID information element is identified by element ID 0. This information element carries the service set identity (SSID) of the IEEE 802.11 WLAN. The length of the SSID may be up to 32 bytes. There is no restriction on the format or content of the SSID. It may be a null-terminated string of ASCII characters or a multibyte binary value. The choice of the value and format of the SSID is entirely up to the network administrator or user. There is one special case for the SSID, when the length of it is zero. In this case, the SSID is considered to be the *broadcast* identity. The broadcast identity is used in Probe Request frames when the mobile STA is attempting to discover all IEEE 802.11 WLANs in its vicinity.

Supported Rates information element

The Supported Rates information element describes the data rates that the STA supports. The element may contain from 1 to 8 bytes of rate information. Each byte represents a single rate, with the 7 LSBs of the byte representing the rate value, and the MSB indicating whether the rate is mandatory. In the base standard, the values describing the rates in this element were mathematically related to the actual data rate. The rates were measured in units of 500 kbit/s. In IEEE 802.11b, the interpretation of the values for the supported rates was changed. The values now are simple labels that are associated with particular data rates. This change was made because of the upper limit of 63.5 Mbit/s the original interpretation imposed on the data rates. With the new interpretation from IEEE 802.11b, there is no upper limit on the data rate imposed by the supported rates element.

The Supported Rates information element is transmitted in Beacon, Probe Response, Association Request, Association Response, Reassociation Request, and Reassociation Response frames. In frames other than the Association Request and Reassociation Request frames, the rates indicated as mandatory must be supported by a STA that desires to be associated with a particular BSS. If a STA does not support all of the rates indicated to be mandatory, it may not associate with the BSS. If it attempts to associate with the BSS, the response from the AP will indicate a failure to associate, with a status code of 18. Rates in the Supported Rates information element that are not mandatory may be used for communication only if both the sender and receiver support those nonmandatory rates.

Finally, there is no correlation between the rates marked as mandatory in the Supported Rates information element and rates that are mandatory to implement for a particular PHY. It may be desirable to mark only rates that are optional for a PHY to be mandatory in the Supported Rates information element. This action can be taken for several reasons, including reducing bandwidth consumption by management and multicast frames that are sent at the lowest mandatory rate, decreasing effective cell size for a BSS [because higher data rates usually require higher signal-to-noise ratio (SNR)], and excluding older equipment that does not implement the newer PHY options. It is the rates marked as mandatory in this information element (and the ERP Information information element, if present) that are used to determine duration values in frames.

FH Parameter Set information element

The FH Parameter Set information element, as well as the DS Parameter Set, CF Parameter Set, and IBSS Parameter Set information elements, is different from the other information elements in that it is not variable in length. It is an information element rather than a fixed field because its presence in management frames is dependent on whether particular options are implemented. For example, the FH Parameter Set information element is present in Beacon and Probe Response frames only if the PHY being used is the IEEE 802.11 frequency hopping spread spectrum (FHSS) PHY or the IEEE 802.11b PHY with the channel agility option enabled. Otherwise, this information element is not present. It is because of the conditional nature of this information element, and the others mentioned, that it is an information element rather than a fixed field. Making it an information element allows the fixed fields of the management frames to be invariant with respect to the PHY or options in use.

The FH Parameter Set information element is 7 bytes long. In addition to the two bytes for the element ID and length, the element contains the dwell time, hop set, hop pattern, and hop index. The description of these items can be found in Chapter 6.

DS Parameter Set information element

The DS Parameter Set information element is 3 bytes long. In addition to the element ID and length, it contains the current channel. This information element is another fixed-length information element, defined as an information element rather than a fixed field because it is present in Beacon and Probe Response frames only if the IEEE 802.11 DSSS or IEEE 802.11b PHY is being used. The description of these items can be found in Chapter 6.

CF Parameter Set information element

The CF Parameter Set information element is 8 bytes long. In addition to the element ID and length, this element contains the CFP count, CFP, CFP maximum duration, and CFP duration remaining. This information element is another fixed-length information element, defined as an information element rather than a fixed field because it is present in Beacon and Probe Response frames only if a PC is in operation in the BSS.

TIM information element

The TIM information element may be from 6 to 256 bytes long. This information element carries information about frames that are buffered at the AP for STAs in power-saving modes of operation. For a description of how the traffic indication map (TIM) is used, see Chapter 9. In addition to the element ID and the length fields, the TIM information element contains four more fields: Delivery TIM (DTIM) Count, DTIM Period, Bitmap Control, and Partial Virtual Bitmap.

The DTIM Count and DTIM Period fields are used to inform mobile STAs when multicast frames that have been buffered at the AP will be delivered and how often that delivery will occur. The AP will buffer all multicast traffic when any mobile STAs are operating in low power modes. (See Chapter 9 for further information.) The DTIM count is an integer value that counts down to zero. This value represents the number of Beacon frames that will occur before the delivery of multicast frames. When the DTIM count is zero, multicast traffic will be sent. The DTIM period is the number of Beacon frames between multicast frame deliveries. The DTIM period has a significant effect on the maximum power savings a STA may achieve if the STA is to receive multicast traffic. The larger the DTIM period, the greater the power savings may be achieved, since a STA may spend a larger proportion of its time in a low power state. Of course, the DTIM period must be balanced with the overall performance of protocols that depend on multicast traffic, as a larger DTIM period will increase the delay before multicast frames are delivered to all STAs.

The Bitmap Control and Partial Virtual Bitmap fields are used to provide information to STAs that have been operating in low power modes about frames that are buffered at the AP. When each STA associates, it is assigned an AID. The value of the AID designates the individual bit in the partial virtual bitmap that, when set, indicates that there is at least one frame buffered at the AP for that STA.

The entire bitmap is 2008 bits long and is not transmitted with each Beacon frame. Only the portion of the bitmap that is necessary to inform STAs of buffered frames is sent. This partial transmission is accomplished by using the Bitmap Control field to identify the starting point of the partial bitmap that is transmitted. The starting point of the partial bitmap is the first byte of the complete bitmap that is nonzero. The ending point of the partial bitmap is the last byte of the complete bitmap that is nonzero.

Seven bits are used in the Bitmap Control field to represent the starting point of the partial virtual bitmap. The value of these 7 bits is N1/2, where N1 is the largest even number so that bits numbered 1 through $(N1 \times 8) - 1$ in the bitmap are all zero. The endpoint of the partial virtual bitmap is represented by N2, where N2 is the smallest number so that bits numbered $(N2 + 1) \times 8$ through 2007 in the bitmap are all zero. The value of the Length field of the TIM information element is set to $(N2 - N1) + 4$.

There is one special case: AID 0. AID 0 is defined to represent buffered multicast frames and is never assigned to an associating STA. The bit representing AID 0 is not used in the virtual bitmap and is always zero. The bit representing AID 0 is carried in the Bitmap Control field.

IBSS Parameter Set information element

The IBSS Parameter Set information element is another fixed-length information element, defined as an information element rather than a fixed field because it will occur in Beacon frames only in an IBSS. In addition to the element ID and length, there is one more field in this information element: Announcement Traffic Indication Message (ATIM) Window. The ATIM Window field is 16 bits long and indicates the length of the ATIM window after each Beacon frame transmission in an IBSS. The length of the ATIM window is indicated in TU.

Challenge Text information element

The Challenge Text information element may be up to 255 bytes long. In addition to the element ID and length fields, this information element carries one more field: Challenge Text. The Challenge Text field may be up to 253 bytes long. There are no format or content restrictions on the Challenge Text field.

Country information element

The Country information element was first defined by IEEE 802.11d to allow the communication of the regulatory domain and restrictions between the STAs in an IEEE 802.11 WLAN. The purpose of this information element was expanded by IEEE 802.11j to provide additional regulatory information that was required to operate in a compliant fashion in the new Japanese bands below 5 GHz. The details of this information element are described in Chapter 7. The changes made to it by IEEE 802.11j are described in Chapter 13.

Hopping Pattern Parameters information element

The Hopping Pattern Parameters information element is defined by IEEE 802.11d to allow for the creation of useful hopping patterns without the need to change the standard. The hopping pattern parameters described in this information element are applied to a formula described in IEEE 802.11d to arrive at a set of hopping pattern tables that are used by an FH PHY. For details on the content of this information element, see Chapter 7.

Hopping Pattern Table information element

The Hopping Pattern Table information element is defined in IEEE 802.11d to allow the communication of a specific hopping pattern between STAs. This information element is provided for cases where an algorithmically derived hopping pattern would not meet regulatory requirements or where more efficient use of a band might be achieved by a tweaked hopping pattern compared to an algorithmically derived pattern. The details of this information element are provided in Chapter 7.

Request information element

The Request information element was defined in IEEE 802.11d to allow a STA to request specific information elements that might not be normally returned in a Probe Response frame. The first need for this capability was seen with the Hopping Pattern Table information element, which might be very large and not desirable to include as a matter of course in the Beacon and Probe Response frames. The Request information element allows any information element to be requested by a STA. See Chapter 7 for details on this information element.

QoS BSS (QBSS) Load information element

The QBSS Load information element is defined in IEEE P802.11e. This information element includes a count of the number of STAs currently associated at an AP, a measure of the channel utilization, and a measure of additional load that the AP can accept. The STA Count field is a simple 16-bit integer that indicates the number of STAs associated at the AP. The Channel Utilization field is 1 byte long and acts as a thermometer showing the amount of that the AP detected the channel to be busy, through either the logical or physical carrier sensing mechanisms. The value of the Channel Utilization field is determined by a formula that works out, basically, to be $255 \times$ (time in microseconds medium is busy during monitor interval) / (monitor interval in microseconds). The available admission capacity is a 16-bit integer that indicates the number of 32 microsecond periods that are available each second. The available admission capacity does not provide a guarantee to a STA that it will be able to use that available capacity.

EDCA Parameter Set information element

The EDCA Parameter Set information element is defined in IEEE P802.11e. It provides a description of the QoS parameters in use at the AP. The QoS Info field provides a 4-bit count that is incremented each time the parameters in the remainder of the information element are changed. This field also indicates the ability of the AP to process QoS acknowledgments, nonzero queue depth requests, and nonzero TXOP limit requests. This field is followed by a single reserved byte and four access category descriptors. Each of the four access category descriptors contains the following information: The first field of the access category descriptor is the ACI/AIFS field. This field contains 4 bits to indicate the number of slots for the AIFS, 2 bits to indicate the access category being described by the descriptor, and 1 bit to indicate when admission control is mandatory for this access category. The second field of the descriptor describes the CW parameters for this access category. This field includes 4 bits each for the CWmin and CWmax values. The actual values for CWmin and CWmax are determined by raising 2 to the power of the value in the 4-bit subfield and subtracting 1 from the result. The third field in the descriptor is the TXOP Limit field. This field indicates the maximum number of 32 μs intervals that are allowed for each transmission in the access category being described. A value of zero in this field indicates that a single frame of any length and data rate, along with any RTS, CTS, and ACK frames, can be transmitted.

TSPEC information element

The TSPEC information element is defined in IEEE P802.11e. This information element is used to describe a specific traffic flow. This information element contains 16 fields: TSINFO, Nominal MSDU Size, Maximum MSDU Size, Minimum Service Interval, Maximum Service Interval, Inactivity Interval, Suspension Interval, Service Start, Minimum Data Rate, Mean Data Rate, Peak Data Rate, Burst Size, Delay Bound, Minimum PHY Rate, Surplus Bandwidth, and Medium Time.

The TSINFO field is 3 bytes long and contains 10 subfields. The Traffic Type subfield indicates whether the traffic flow described by this TSPEC is periodic. The TSID subfield, along with the direction and MAC address, uniquely identifies the flow. This subfield is 4 bits long. The 2-bit Direction subfield indicates if the flow direction is uplink (STA to AP), downlink (AP to STA), bidirectional (both STA to AP and AP to STA), or direct link (STA to STA). The Access Policy subfield describes the types of access that will be used by the flow. The Aggregation subfield is 1 bit long. It is used to indicate when an aggregate schedule is being requested for the flow by the STA or when an aggregate schedule is being provided by the AP. This bit may be set only if the access policy is hybrid coordinated channel access (HCCA) or if the access policy is enhanced distributed channel access (EDCA) and the traffic type is periodic. The APSD subfield is 1 bit. It is used to indicate when the associated flow will use automatic power save delivery (APSD). The User Priority subfield is a 3-bit field that indicates the IEEE 802.1D™ priority value for the flow. The ACK Policy subfield indicates the type of acknowledgment that is to be used for this flow. The allowable values for this subfield indicate normal ACK, block ACK, or no Ack is to be used. The Schedule subfield is used only when the access policy is EDCA and reserved otherwise. The remainder of the field is reserved.

The Nominal MSDU Size field is 16 bits long and indicates the MSDU size and if the MSDU size is fixed or variable. When the MSDU size is fixed, the field indicates the size of the MSDU. When the MSDU size is variable, the field indicates the nominal size of the MSDU. The term *nominal* is not defined in the IEEE 802.11 standard. It could mean the average size of the MSDU or the size that is used most often for the MSDU. For the purposes of allowing the most accurate calculation of channel utilization, it is recommended that the average MSDU size be used for this field when the size is variable. The value of zero for MSDU size, when the MSDU size is variable, means that the MSDU size is unspecified.

The Maximum MSDU Size field is 16 bits long and indicates the maximum number of bytes in an MSDU in the flow being described.

The Minimum Service Interval field is 32 bits long and indicates the minimum number of microseconds between the start of two successive service periods for the flow being described.

The Maximum Service Interval field is 32 bits long and indicates the maximum number of microseconds between the start of two successive service periods for the flow being described.

The Inactivity Interval field is 32 bits long and indicates the minimum number of microseconds that must elapse without any activity on the flow being described before the flow will be deleted.

The Suspension Interval field is 32 bits long. The value of this field must be less than the value of the inactivity interval and indicates the minimum number of microseconds that must elapse before the AP may suspend polling the STA for frames belonging to the flow being described.

The Service Start field is 32 bits long and is used only when the APSD subfield of the TSINFO field is nonzero. When the APSD subfield is zero, the Service Start field is also zero. When the APSD subfield is nonzero, the Service Start field indicates the 32 LSBs of the timer synchronization function (TSF), at which time the STA will be ready to begin transferring frames belonging to the flow being described. This mechanism can be used for scheduling algorithms to minimize the time that a STA must remain awake and waiting for frame delivery from the AP.

The Minimum Data Rate field is 32 bits long and specifies the lowest data rate at which frames belonging to the flow being described will be transferred. This field is in units of bits per second.

The Mean Data Rate field is 32 bits long and specifies the average data rate used to transfer any frames belonging to the flow being described. This field is in units of bits per second.

The Peak Data Rate field is 32 bits long and specifies the peak rate at which frames belonging to the flow being described will be transferred. This field is in units of bits per second. During no interval of time may frames be transferred that exceed the value of this field.

The Burst Size field is 32 bits long and specifies the maximum number of bytes that will be required to be transferred at the peak data rate. If this field is zero, there are no bursts.

The Delay Bound field is 32 bits long and specifies the upper bound, in microseconds, for delay of any single frame belonging to the flow being described. This delay includes all MAC layer queueing, MAC retransmissions, and MAC acknowledgment.

The Minimum PHY Rate field is 32 bits long and specifies the lowest data rate, in bits per second, to be used to transmit frames belonging to the flow being described.

The Surplus Bandwidth field is 16 bits long and is a fixed point value, with 3 bits in the integer portion and 13 bits in the fractional portion. The value of this field indicates the ratio of additional bandwidth that should be allocated to the flow being described to allow for retransmissions, even at the minimum PHY rate, in order to meet the throughput and delay bounds for the flow being described. This field should always have a value greater than or equal to 1.0.

The Medium Time field is a 16-bit integer, specifying the number of 32 µs blocks of time in each second that are allocated to the flow being described. This field is reserved in the information element in the ADDTS action frame.

TCLAS information element

The TCLAS information element is used to allow the classification of frames and assignment to an individual flow that has been described in a TSPEC information element. The IEEE 802.11 standard is clear that this information element can be used with downlink and bidirectional links. It seems ambiguous about whether it is to be used with uplink or direct link traffic, using the wording "need not be provided." It is recommended that this information element is not used for uplink and direct link traffic because the source of the traffic should have sufficient internal information to classify the traffic and assign it to a flow.

The TCLAS information element has two fields: User Priority and Frame Classifier.

The User Priority field is 1 byte long and includes the 3-bit IEEE 802.1D user priority value for the frames associated with this flow.

The Frame Classifier field has three subfields: Classifier Type, Classifier Mask, and Classifier Parameters. The Classifier Type subfield indicates that the remainder of the field is describing the method to classify frames according to Ethernet parameters, TCP/UDP IP parameters, or IEEE 802.1D/Q parameters. The Classifier Mask subfield is bit significant and indicates which of the values in the following Classifier Parameters subfield are required to match the equivalent fields in a frame in order to assign the frame to an individual flow. The IEEE 802.11 standard is not clear on how the bits in the classifier mask map to the individual fields in the classifier parameters. It is expected that the mapping is for bit 0 of the mask to apply to the first parameter in the classifier parameters, bit 1 to apply to the second parameter, bit 2 to the third parameter, and so on.

The Classifier Parameters subfield has a format that depends on the classifier type value. For a classifier type value of 0 (Ethernet), the classifier parameters are shown in Table 3–8. Classifier parameters for TCP/UDP IP are shown in Table 3–9 and Table 3–10. Classifier parameters for IEEE 802.1D/Q are show in Table 3–11.

Table 3–8:   Ethernet frame classifier parameters

Classifier mask bit	Parameter
0	Destination address
1	Source address
2	Ethertype

Table 3–9:   TCP/UDP IP classifier parameters, for IPv4

Classifier mask bit	Parameter
0	Version
1	Source IP address
2	Destination IP address
3	Source port
4	Destination port
5	DSCP
6	Protocol

Table 3–10:   TCP/UDP IP classifier parameters, for IPv6

Classifier mask bit	Parameter
0	Version
1	Source IP address
2	Destination IP address
3	Source port
4	Destination port
5	Flow label

Table 3–11:   IEEE 802.1D/Q classifier parameters

Classifier mask bit	Parameter
0	IEEE 802.1Q^{TM} tag type

Schedule information element

The Schedule information element is defined in IEEE P802.11e. It is used by an AP to provide schedule information to a STA for an individual traffic flow. The information element contains four fields: Schedule Info, Service Start Time, Service Interval, and Specification Interval.

The Schedule Info field is 16 bits long. It contains three subfields: Aggregation, TSID, and Direction. The Aggregation subfield is 1 bit long and indicates if the schedule described in the information element is an aggregate of all traffic flows for the STA or is for a single traffic flow. When the value of the Aggregation subfield is one, the schedule information is for all traffic flows for the STA. When the value of the Aggregation subfield is zero, the schedule information is for a single traffic flow. The TSID and Direction subfields indicate the traffic flow to which the schedule applies when the value of the Aggregation subfield is zero. These subfields are reserved when the value of the Aggregation subfield is one. The remaining 8 bits of the Schedule Info field are reserved.

The Service Start Time field is a 32-bit integer indicating the time, expressed as the value of the 32 LSBs of the TSF timer, when the service is anticipated to begin.

The Service Interval field is a 32-bit integer indicating the time, in microseconds, between two successive service periods.

The Specification Interval field is a 16-bit integer indicating the time, in TUs, over which schedule compliance will be measured.

Power Constraint information element

The Power Constraint information element is defined in IEEE 802.11h. For details on this information element, see Chapter 6.

Power Capability information element

The Power Capability information element is defined in IEEE 802.11h. For details on this information element, see Chapter 6.

TPC Request information element

The TPC Request information element is defined in IEEE 802.11h. For details on this information element, see Chapter 6.

TPC Report information element

The TPC Report information element is defined in IEEE 802.11h. For details on this information element, see Chapter 6.

Supported Channels information element

The Supported Channels information element is defined in IEEE 802.11h. For details on this information element, see Chapter 6.

Channel Switch Announcement information element

The Channel Switch Announcement information element is defined in IEEE 802.11h. For details on this information element, see Chapter 6.

Measurement Request information element

The Measurement Request information element is defined in IEEE 802.11h. For details on this information element, see Chapter 6.

Measurement Report information element

The Measurement Report information element is defined in IEEE 802.11h. For details on this information element, see Chapter 6.

Quiet information element

The Quiet information element is defined in IEEE 802.11h. For details on this information element, see Chapter 6.

IBSS DFS information element

The IBSS DFS information element is defined in IEEE 802.11h. For details on this information element, see Chapter 6.

ERP Information information element

The ERP Information information element is defined in IEEE 802.11g. This information element is present in Beacon and Probe Response frames. It contains a single, bit-significant field. Bit 0 of the field indicates that a non-ERP (i.e., IEEE 802.11g) STA is present in the BSS. This setting may indicate that the non-ERP STA is associated with the AP sending this information element or that there is an overlapping BSS (on the same channel) with non-ERP STAs present. Bit 1 of the field indicates that STAs associating with the AP sending the information element must use a protection method with all transmissions in the BSS. Bit 2 of the field indicates that Barker-mode preamble is required to be used by STAs associated with the AP sending the information element. When this bit is not set, the Barker-mode preamble is not required. Bits 3 through 7 of the field are reserved.

TS Delay information element

The TS Delay information element is defined by IEEE P802.11e. This information element is sent by a STA after denying an ADDTS request. The delay field is a 32-bit integer indicating the amount of time, measured in TUs, that the receiving STA is being told by the AP to wait before requesting the addition of a traffic flow. If the value of this field is zero, the STA is being told to wait an indeterminate amount of time before requesting the addition of a new traffic flow.

TCLAS Processing information element

The TCLAS Processing information element is defined by IEEE P802.11e. This information element is used to qualify how TCLAS information elements in the same frame are to be applied to downlink and bidirectional flows. This information element has a single, 1-byte integer processing field. The value of the processing field has three defined values. When the value of the processing field is zero, all TCLAS elements must match a frame to assign it to the flow. When the value of the processing field is one, a match of one or more TCLAS elements assign a frame to the flow. When the value of the processing field is two, any frames that are not assigned to another flow are assigned to the flow identified in the frame containing the TCLAS Processing information element. There should be no TCLAS information elements in the frame when the value of the processing field of this information element is two. All other values of the processing field are reserved.

QoS Capability information element

The QoS Capability information element is defined in IEEE P802.11e. This information element is sent in Beacon frames that do not include an EDCA Parameters information element and in Reassociation Request frames. There is a single, 1-byte QoS Info field in this information element. The format of this field is identical to the field of the same name in the EDCA Parameters information element.

RSN information element

The RSN information element is defined in IEEE 802.11i. For a detailed description of this information element, see Chapter 4.

Extended Supported Rates information element

The Extended Supported Rates information element is defined in IEEE 802.11g. This information element is necessary because of the limitation of the Supported Rates information element to carry only information on eight data rates. With the adoption of IEEE 802.11g, it became possible to support more than eight data rates. The Extended Supported Rates information element has the same format as the Supported Rates information element, without the limitation on the number of rates that may be included in the information element.

Vendor-Specific information element

The Vendor-Specific information element was adopted by the IEEE 802.11 Working Group by motion and will be incorporated in the standard with an amendment that is targeted for publication in 2005. The purpose of this information element is two-fold. First, the Vendor-Specific information element is to stop the proliferation of uncoordinated usurpation of information element IDs that are reserved by the standard for future standardization. Second, the Vendor-Specific information element standardizes some of the information in the element. The Vendor-Specific information element includes two fields: Organizational Unique Identifier (OUI) and Vendor-Specific Content. The OUI field contains the 3-byte, IEEE-assigned OUI of the vendor that defined the content of the information element in the same order that the OUI would be transmitted in an IEEE 802.11 address field. The Vendor-Specific Content field contains any information the identified vendor chooses, in any format.

OTHER MAC OPERATIONS

Fragmentation

The wireless medium on which the IEEE 802.11 WLAN operates is unlike wired media in many ways. One of the most significant ways that it differs from wired media is the presence of uncontrollable interference, particularly in the 2.4 GHz radio band, that can render communication between WLAN STAs nearly impossible. In addition to other communication users of the 2.4 GHz band, this band is also a worldwide industrial, scientific, medical (ISM) band. Probably the most widespread equipment using this band is the conventional microwave oven.

A microwave oven produces its microwave radiation using a magnetron. To keep the cost of the oven to a minimum, the magnetron is typically powered by a half-wave rectified power supply. This supply system causes the magnetron to be emitting radiation through half of each 60 Hz power cycle (in the United States). In other words, the magnetron is radiating for 8 ms of every 16 ms. In addition, the frequency of operation of the magnetron changes as the power supply ramps up and down, sweeping the radiation across a large portion of the band.

To operate in the presence of this known interferer, the IEEE 802.11 MAC can fragment its frames in an attempt to increase the probability that they will be delivered without errors induced by the interference. Frames longer than the fragmentation threshold are fragmented prior to the initial transmission into fragments no longer than the value of the dot11FragmentationThreshold attribute in the MIB. A frame will be divided into one or more fragments equal in length to the fragmentation threshold and no more than one fragment smaller than the fragmentation threshold. The default value of the fragmentation threshold is such that no frames will be fragmented. The value may be changed to begin fragmenting frames. If the interference source is known, such as the microwave oven, the value of the fragmentation threshold may be calculated from the characteristics of the interferer and the bit rate of the transmissions.

When a frame is fragmented, the Sequence Control field of the frame header indicates the placement of the individual fragment among the set of fragments. The More Fragments subfield in the Frame Control field indicates whether the current fragment is the last fragment. The fragments are transmitted in order by their fragment numbers in the Sequence Control field. The lowest numbered fragment is transmitted first. Subsequent fragments are transmitted immediately upon receiving the acknowledgment of the previous fragment, without needing to compete for the medium again. This process is called a *fragment burst*. IEEE 802.11 uses fragment bursts to minimize the total amount of time that is taken to deliver a single frame that has been fragmented, when all of the fragments are delivered on the first attempt. If a fragment is not delivered on the first attempt, subsequent fragments are not transmitted until the previous fragment has been acknowledged. If a fragment is not acknowledged, the normal rules for retransmission of frames apply. See Figure 3–13.

The duration calculation during a fragment burst is slightly different from that of the calculation during a single frame transmission. For a fragment burst, the calculation of duration for the RTS and CTS frames, if any, is unchanged. The duration calculated for the data and ACK frames changes, however.

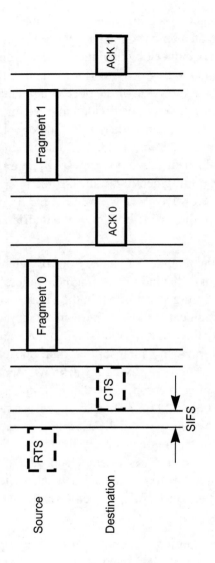

Figure 3–13:   Fragmentation of a data frame

Normally, the duration value in a data frame is calculated from the end of the data frame to the end of the ACK frame that is expected to follow. In a fragment burst, the duration value in a data frame is similar to that of the RTS frame and is calculated from the end of the current fragment to the end of the subsequent fragment, including the ACK frame that is interposed. Similarly, the duration calculated for an ACK frame in a fragment burst is like the duration of the CTS frame. It is calculated from the end of the current ACK frame to the end of the subsequent ACK frame and includes the data frame that is interposed. This modified calculation of the duration for data and ACK frames is used throughout the fragment burst, except for the last data frame, the one that has the More Fragments subfield clear, and the final ACK frame. For these two final frames of a fragment burst, the duration calculation reverts to the normal duration calculation. The duration of the final data frame will be calculated from the end of the data frame to the end of the ACK frame. The value of the duration in the final ACK frame is zero. See Figure 3–14.

Privacy

It should be noted that the use of the wired equivalent privacy (WEP) mechanism is not recommended, due to the several successful, practical, and theoretical attacks that have been published. The material in this section is provided for completeness.

IEEE 802.11 incorporates MAC level privacy mechanisms to protect the content of data frames from eavesdropping. This need is, again, because the medium for the IEEE 802.11 WLAN is significantly different from that of a wired LAN. The WLAN lacks even the minimal privacy provided by a wired LAN. The wired LAN must be physically compromised in order to tap its data. A WLAN, by contrast, can be compromised by anyone with a suitable antenna. The IEEE 802.11 WEP mechanism provides protection at a level that is felt to be equivalent to that of a wired LAN.

WEP is an encryption mechanism that takes the content of a data frame, its frame body, and passes it through an encryption algorithm. The result then replaces the frame body of the data frame and is transmitted. Data frames that are encrypted are sent with the Protected Frame bit (formerly WEP bit) in the Frame Control field of the MAC header set. The receiver of an encrypted data frame passes the encrypted frame body through the same encryption

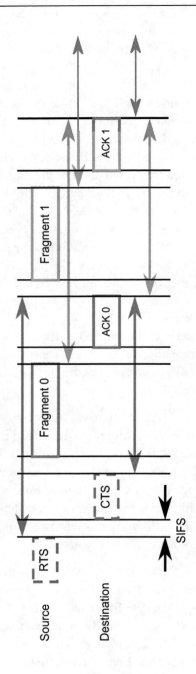

Figure 3–14: NAV setting during fragmentation

algorithm used by the sending STA. The result is the original, unencrypted frame body. The receiver then passes the unencrypted result up to higher layer protocols.

It should be noted that only the frame body of data frames is encrypted. This limitation leaves the complete MAC header of the data frame, and the entire frame of other frame types, unencrypted and available to even the casual eavesdropper. Thus, WEP does provide protection for the content of the data frames, but does not protect against other security threats to a LAN, such as traffic analysis.

The encryption algorithm used in IEEE 802.11 is RC4. RC4 was developed by Dr. Ron Rivest of RSA Data Security, Inc. (RSADSI). RSADSI is now part of Network Associates, Inc. RC4 is a symmetric stream cipher that supports a variable-length key. A symmetric cipher is one that uses the same key and algorithm for both encryption and decryption. A stream cipher is an algorithm that can process an arbitrary number of bytes. This cipher is contrasted with a block cipher that processes a fixed number of bytes. The key is the one piece of information that must be shared by both the encrypting and decrypting STAs. It is the key that allows every STA to use the same algorithm, but only those STAs sharing the same key can correctly decrypt encrypted frames. RC4 allows the key length to be variable, up to 256 bytes, as opposed to requiring the key to be fixed at a certain length. IEEE 802.11 has chosen to use a 40-bit key.

The IEEE 802.11 standard describes the use of the RC4 algorithm and the key in WEP. However, key distribution or key negotiation is not mentioned in the standard. This omission leaves much of the most difficult part of secure communications to the individual manufacturers of IEEE 802.11 equipment. In a secure communication system using a symmetric algorithm, such as RC4, it is imperative that the keys used by the algorithm be protected, i.e., that they remain secret. If a key is compromised, all frames encrypted with that key are also compromised. Thus, while it is likely that equipment from many manufacturers will be able to interoperate and exchange encrypted frames, it is unlikely that a single mechanism will be available that will securely place the keys in the individual STAs. There is currently discussion in the IEEE 802.11 Working Group to address this lack of standardization.

WEP details

IEEE 802.11 provides two mechanisms to select a key for use when encrypting or decrypting a frame. The first mechanism is a set of as many as four default keys. Default keys are intended to be shared by all STAs in a BSS or an ESS. The benefit of using a default key is that, once the STA obtains the default keys, a STA can communicate securely with all of the other STAs in a BSS or ESS. The problem with using default keys is that they are widely distributed to many STAs and may be more likely to be revealed. The second mechanism provided by IEEE 802.11 allows a STA to establish a *key mapping* relationship with another STA. Key mapping allows a STA to create a key that is used with only one other STA. Although this one-to-one mapping is not a requirement of the standard, this method would be the most secure way for a STA to operate because there would be only one other STA that would have knowledge of each key used. The fewer STAs possessing a key, the less likely the key will be revealed.

The dot11PrivacyInvoked attribute controls the use of WEP in a STA. If dot11PirvacyInvoked is false, all frames are sent without encryption. If dot11PrivacyInvoked is true, all frames will be sent with encryption, unless encryption is disabled for specific destinations. Encryption for specific destinations may be disabled only if a key mapping relationship exists with that destination. The dot11PrivacyInvoked attribute controls only the use of WEP for transmission. On reception the Protected Frame bit (formerly WEP bit) controls whether the receiving STA must have a key and attempt to decrypt the received frame.

A default key may be used to encrypt a frame only when a key mapping relationship does not exist between the sending and receiving STA. When a frame is to be sent using a default key, the STA determines whether any default keys are available. Four possible default keys might be available. A key is available if its entry in the dot11WEPDefaultKeysTable is not null. If at least one default key is available, the STA chooses one key, by an algorithm not defined in the standard, and uses it to encrypt the frame body of the frame to be sent. The WEP header and trailer are appended to the encrypted frame body. The default key used to encrypt the frame is indicated in the KeyID of the header portion along with the initialization vector, and the integrity check value (ICV) is indicated in the trailer. If there are no available default keys, i.e., all default keys are null, the frame is discarded. See Figure 3–15.

If a key mapping relationship exists between the source and destination STAs, the *key mapping key*, the key shared only by the source and destination STAs, must be used to encrypt frames sent to that destination. When a frame is to be sent using a key mapping key (i.e., the dot11WEPKeyMappingWEPOn entry for the destination is true), the key corresponding to the destination of the frame is chosen from the dot11WEPKeyMappingsTable. The frame body is encrypted using the key mapping key, and the WEP header and trailer are appended to the encrypted frame body. The value of the KeyID is set to zero when a key mapping key is used. If the value of dot11WEPKeyMapping-WEPOn for the destination is false, the frame is sent without encryption.

Corresponding to the dot11PrivacyInvoked attribute controlling the sending of frames, the dot11ExcludeUnecrypted attribute controls the reception of encrypted frames. When dot11ExcludeUnecrypted is false, all frames addressed to the STA are received, regardless of whether they are encrypted. However, when dot11ExcludeUnecrypted is true, the STA will receive only frames that are encrypted and discard all data frames that are not encrypted. If a frame is discarded because it is not encrypted and dot11ExcludeUnecrypted is true, there is no indication to the higher layer protocols that any frame was received.

Two counters are associated with WEP. The dot11UndecryptableCount attribute reflects the number of encrypted frames that were received by the STA that could not be decrypted, either because a corresponding key did not exist or because the WEP option is not implemented. The dot11ICVError-Count attribute reflects the number of frames that were received by a STA for which a key was found that should have decrypted the frame, but that resulted in the calculated ICV value not matching the ICV received with the frame. These two counters should be monitored carefully when WEP is used in a WLAN. A rapidly increasing dot11UndecryptableCount attribute can indicate that an attack to deny service might be in progress. A rapidly increasing dot11ICVErrorCount attribute can indicate that an attack to determine a key is in progress.

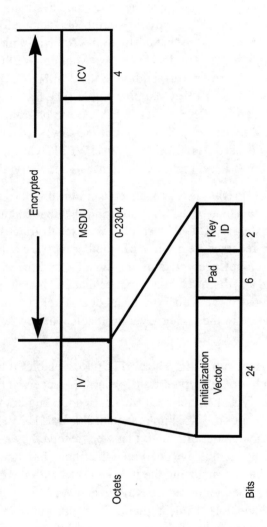

Figure 3–15: WEP expansion of the frame body

Chapter 4   IEEE 802.11i security enhancements

Late in 1999 and continuing into 2000, several authors published research on weaknesses of the wired equivalent privacy (WEP) algorithm of the IEEE 802.11 base standard. Shortly after these results were published, attack programs began to appear on the Internet that allowed anyone to attempt to break the mechanism used to protect a WLAN or to inject packets into a WLAN without detection. This development posed a significant problem to WLAN vendors who were just beginning to see the wider acceptance of IEEE 802.11, particularly IEEE 802.11b, by commercial enterprises and the government. The newly published security vulnerabilities of IEEE 802.11 caused nearly all of that interest in the enterprise and government markets to evaporate.

In March 2000, the project for IEEE 802.11 Task Group e (TGe) was approved. Originally, TGe was to work on both security enhancements and QoS enhancements. In May 2001, Task Group i (TGi) was split from TGe to focus only on security enhancements. The task for TGi was two-fold: to create a new, very secure means of authentication and privacy that would not be vulnerable to the weaknesses of WEP and to, somehow, "fix WEP." An amendment to the IEEE 802.11 standard, IEEE 802.11i, was approved in 2004. The result of this work is the robust security network (RSN) and the transition security network (TSN). A RSN is a WLAN using one of the cipher suites defined in IEEE 802.11i [i.e., Temporal Key Integrity Protocol (TKIP) or Counter with CBC-MAC Protocol (CCMP)] for both the pairwise and group ciphers. A TSN is a WLAN that is using WEP for the group cipher.

ROBUST SECURITY NETWORK (RSN)

TGi has developed the RSN specification to provide very strong confidentiality (i.e., encryption) and strong per-packet authenticity. RSN addresses the shortcomings of WEP in these areas with new encryption algorithms (called *cipher suites* in IEEE 802.11i) and new integrity mechanisms. It needs to be noted that IEEE 802.11i does not address all of the security problems that plague a WLAN. IEEE 802.11i specifically addresses authentication and confidentiality of data frames. It does not address other shortcomings of IEEE 802.11, such as protection of management frames, prevention of denial of service attacks, or prevention of attacks that may take place in higher layers, e.g., ARP spoofing. This section describes the general characteristics of an RSN WLAN.

An RSN WLAN is identified by the presence of a new information element in the Probe Response and Beacon frames. The new information element is the RSN information element. The content of this information element identifies the group key cipher suite, the pairwise key cipher suite, the authentication and key management (AKM) suite, and other capabilities of the RSN WLAN. If the group key cipher suite is WEP in the RSN information element, the WLAN is a TSN.

TGi introduced a negotiation between STAs in RSN WLANs to determine the cipher suites and AKM suite to be used between the two STAs. While the IEEE 802.11 standard allows any two STAs to perform the negotiation, the predominant usage is between a STA and an AP. Therefore, the remainder of the description in this section uses a STA and an AP as the example. Figure 4–1 shows the protocol handshake described by IEEE 802.11i.

The negotiation begins with the announcement of the available cipher suites and AKM suites in the RSN information element in the Beacon and Probe Response frames from the AP. IEEE 802.11i requires that the Protected Frame bit (formerly the WEP bit) in the Capability Information field must be set to one to indicate that the WLAN identified by the SSID requires the use of one of the cipher suites in the RSN information element to protect the content of data frames.

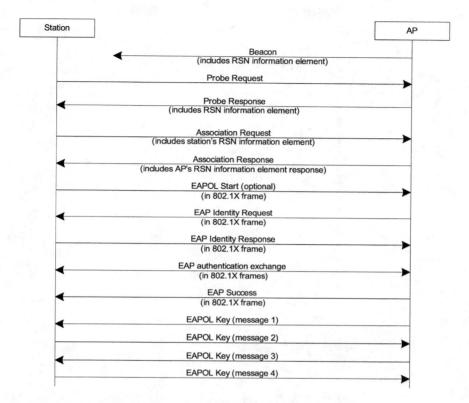

Figure 4–1: IEEE 802.11i protocol handshake

When a STA associates, it places an RSN information element (see Figure 4–2) in the Association Request frame with the values for the cipher suites and AKM suites that it has selected from among the values from the corresponding information element in the Beacon or Probe Response frame that are also supported by the STA. IEEE 802.11i also requires the Protected Frame bit in the Capability Information field of the Association Request frame to be set. In the Association Response frame, the STA receives the values from the AP that are to be used for the duration of the association if the association is successful, or it receives a status code that indicates the exact failure that must be remedied for the association to succeed. The particular failures that are indicated by new status code values are invalid group cipher, invalid pairwise cipher, invalid Authentication and Key Management Protocol

Element ID	Length	Version	Group Cipher Suite	Pairwise Cipher Suite (m)	Pairwise Cipher Suite List	AKM Suite Count (n)	AKM Suite List	RSN Capabilities	PMKID Count (s)	PMKID List
Octet: 1	1	2	4	2	4m	2	4n	2	2	16s

Figure 4–2:   RSN information element

(AKMP), invalid version in the information element, invalid capabilities in the information element, and failure due to security policy rejection. Once more, IEEE 802.11i requires the Protected Frame bit in the Capability Information field to be set in the Association Response frame. While the association response is not authenticated, its content is validated later in the 4-Way Handshake.

The value of the Element ID field for the RSN information element is 48. The Length field indicates the number of bytes that follow the Length field. For IEEE 802.11i, the value of the Version field is one. All other values of the Version field are reserved for future development of the IEEE 802.11 standard.

The Group Key Cipher Suite field encodes the value of the cipher suite that will be used to encrypt group-addressed frames. The field uses the format of a cipher suite selector. The format of cipher suite selector is a 3-byte organizational unique identifier (OUI), assigned by the IEEE, followed by a 1-byte value for the suite type, identifying the particular cipher suite defined by the owner of the OUI. The value of the OUI defined for IEEE 802.11i-defined cipher suites is 00-0f-ac. All other values for the OUI represent definitions of proprietary cipher suites, defined by the owner of the OUI. The values for the suite type are defined for IEEE 802.11i in Table 4–1. The table defines suite types for both the group and pairwise cipher suites.

Table 4–1:   Suite types and their definitions

Suite type	Meaning
0	Use group key cipher suite
1	WEP-40
2	TKIP
3	Reserved
4	CCMP – default in a robust security network association (RSNA)
5	WEP-104
6–255	Reserved

The WEP-40 and WEP-104 values can be used only as group key ciphers. They are never allowed as pairwise ciphers by IEEE 802.11i. Using WEP-40 or WEP-104 as a group key cipher indicates that the WLAN is a TSN.

The value of the Pairwise Cipher Suite Count field indicates the number of pairwise cipher suites that follow the field. The value of this field cannot be zero.

The Pairwise Cipher Suite List field contains a nonempty list of cipher suite selectors.

The value of the AKM Suite Count field indicates the number of AKMPs that follow the field. The value of this field cannot be zero.

The AKM Suite List field contains a nonempty list of AKM suite selectors. The AKM suite selector format is identical to the cipher suite selector format: an OUI followed by a suite selector. The OUI for suite selectors defined by IEEE 802.11i is 00-0f-ac. Table 4–2 defines the values of the suite type for IEEE 802.11i-defined AKM suites.

Table 4–2:   AKM suites and their definitions

Suite type	Meaning	
	Authentication type	**Key management type**
0	Reserved	Reserved
1	Authentication negotiated over IEEE 802.1X or using pairwise master key security association (PMKSA) caching as defined in 8.4.6.2 of the IEEE 802.11 standard – RSNA default	RSNA key management as defined in 8.5 of the IEEE 802.11 standard or using PMKSA caching as defined in 8.4.6.2 of the standard – RSNA default
2	Preshared key (PSK)	RSNA key management as defined in 8.5, using PSK
3–255	Reserved	Reserved

The RSN Capabilities field is 16 bits (2 bytes) in length. Figure 4–3 shows the allocation of the bits in this field.

- The Preauth bit is used only by APs. The value of this bit is always zero when transmitted by a STA. An AP uses this bit to indicate whether it supports IEEE 802.11i preauthentication. When this bit is set, the AP supports IEEE 802.11i preauthentication. When this bit is clear, the AP does not support IEEE 802.11i preauthentication.

  Preauthentication is performed by a STA by communicating to one AP through another AP. The AP with which the STA is attempting to preauthenticate must advertise the Preauth bit in its RSN information element and must also be reachable through the AP with which the STA is currently associated. The term *reachable* generally means that the AP where the STA is attempting to authenticate and the STA must be on the same layer 2 LAN (including VLAN) for the IEEE 802.1X preauthentication frames to be able to be communicated. Reachability is a problem that is not solved by IEEE 802.11. The STA will not have any information provided to it that will indicate whether one AP is reachable through another AP. The STA will have to speculatively try to preauthenticate in order to determine reachability. The STA must also take into account that its preauthentication frames were successfully received by the AP where it is associated when determining that another AP is not reachable. Frame loss may cause a false determination that an AP is not reachable.

  The AP where the STA is attempting to preauthenticate must also recognize its BSSID on its DS interface as that is the only address by which the STA knows the AP. Preauthentication uses the same IEEE 802.1X frame format and state machine as for authentication except that the Ethertype value is changed to 88-c7.

- The No Pairwise bit is used only by STAs. (The value of this bit is always zero when transmitted by an AP.) If this bit is set, the STA does not simultaneously support a WEP default (i.e., shared) key installed at KeyID 0 and a separate pairwise key. If this bit is clear, the STA does support both a WEP default key installed at KeyID 0 and a separate pairwise key.

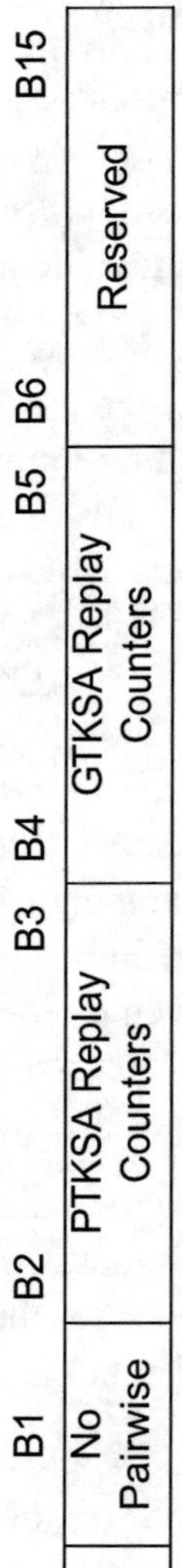

B0	B1	B2		B3	B4		B5	B6		B15
Preauth	No Pairwise	PTKSA Replay Counters			GTKSA Replay Counters			Reserved		

Figure 4–3: RSN Capabilities field

- The PTKSA Replay Counters subfield is 2 bits long and indicates the number of replay counters per pairwise transient key security association (PTKSA) supported by the sender of the RSN information element. The encoding of the field is given in Table 4–3. The intention for having multiple replay counters is that a unique replay counter is needed for each priority level (per IEEE P802.11e), in order to maintain the properties of monotonicity required for the replay algorithm when frames may be reordered before transmission to satisfy priority requirements.

Table 4–3: PTKSA Replay Counter subfield encoding

Number of replay counters value	Meaning
0	1 replay counter
1	2 replay counters
2	4 replay counters
3	16 replay counters

- The GTKSA Replay Counters subfield is 2 bits long and indicates the number of replay counters for the group temporal key security association (GTKSA) supported by the sender of the RSN information element. The encoding is identical to the encoding of the PTKSA Replay Counter subfield given in Table 4–3. The intention for having multiple GTKSA replay counters is also the same as for the PTKSA replay counters: to provide a unique replay counter for each priority level and maintain the properties of the replay algorithm.

- All remaining bits in the RSN Capabilities field are reserved for future standardization.

The value of the PMKID Count field indicates the number of pairwise master key identifiers (PMKIDs) that follows the field. The value of this field may be zero. This field is also optional and may not be present in the RSN information element. Implementations should be prepared to receive RSN information elements where this field is not present, where the value is zero, and where the value is greater than zero.

The PMKID List field is optional. It appears only in an RSN information element in a Reassociation Request frame if the PMKID Count field is present in the same information element and the value of that field is nonzero. The format of the 128-bit PMKID List field equals the first 128 bits of the result of a hashed message authentication code (HMAC) Secure Hash Algorithm 1 (SHA-1) operation. The PMKID is used by the STA to identify a cached PMKSA with the AP. This cached security association can be created through a direct IEEE 802.1X exchange with the AP or preauthentication. It is possible to use the PMKID to identify a cached security association for an AP using a PSK. However, because the purpose of pairwise master key (PMK) caching is to eliminate the Extensible Authentication Protocol (EAP) authentication exchange upon reassociating with an AP and there is no EAP authentication exchange when using PSKs, the usefulness of this configuration is debatable.

A very useful extension to PMK caching has been developed, called *proactive key caching (PKC)*, that has the potential to significantly reduce the number of EAP authentication exchanges in a WLAN, perhaps to only a single EAP exchange at the first AP encountered in the WLAN. PKC requires the AP infrastructure to be able to access the original PMKSA whenever a STA reassociates in the WLAN. To make use of this capability, the STA must speculatively include a PMKID, based on the last PMK obtained in the WLAN and the BSSID of the AP to which it is reassociating, in every Reassociation Request frame. The cost to the STA is minimal, and the benefit can be great. If the PMK is available at the AP, even though the STA has never authenticated or preauthenticated at that AP, PKC eliminates the EAP authentication exchange, saves time during reassociation, and reduces the transaction load on the authentication server as well.

Once the STA is successfully associated, it must then complete authentication using the selected AKMP. This step is a significant deviation from the initial architecture for authentication and association in the IEEE 802.11 base standard, where once association was complete, data frames flow freely between the STA and AP. With IEEE 802.11i, only data frames related to the

AKMP are allowed to be exchanged between the STA and AP. The AP is required to drop all other data frames received from the STA or destined to the STA until the AKMP is completed successfully. This approach can cause problems with higher layers if the AKMP takes a significant amount of time to complete and the medium is indicated to be "connected" during this time. Typically, higher layer packets may be dropped during the interval between completion of the association and completion of the AKMP.

The available protocols for AKMP are the EAP and the PSK. IEEE 802.1X is utilized to encapsulate the authentication exchange between the Supplicant in the STA and the authentication system for EAP, which may be in the AP or in another system connected to the AP, such as a remote authentication dial-in user service (RADIUS) server. PSK requires that a shared secret, the key, must be present in both the AP and STA. Both EAP and PSK utilize IEEE 802.1X to perform the key exchange.

IEEE 802.1X Authentication and Key Management Protocol (AKMP)

Using IEEE 802.1X, the AP performs the role of the authenticator and implements the IEEE 802.1X and EAP state machines, as well as the IEEE 802.1X controlled and uncontrolled ports. The IEEE 802.1X authenticator receives EAP packets in IEEE 802.1X frames from the IEEE 802.1X supplicant in the STA and forwards those EAP packets to the authentication system, typically a RADIUS server. The uncontrolled port is used to pass only IEEE 802.1X frames between the STA's supplicant and the AP's authenticator. The controlled port is closed by the IEEE 802.1X state machine until the successful completion of the EAP authentication. This closure prevents the passage of any frames on the controlled port during the authentication process. In effect, the AP allows only IEEE 802.1X frames between itself and the STA, over the uncontrolled port, until the authentication system informs the AP that the authentication has completed successfully. Generally, encryption is not applied to IEEE 802.1X frames. The protection of the information in the authentication protocol is expected to be provided by the authentication protocol itself. However, IEEE 802.1X frames are carried in normal IEEE 802.11 data frames. Therefore, if keys are present at both the AP and STA from a previous authentication exchange, the

IEEE 802.1X frame exchange may be encrypted with those keys. Upon successful authentication, the AP opens the controlled port and all traffic to and from the STA is allowed.

At the end of an IEEE 802.1X authentication exchange, the STA and the authentication system have developed a shared PMK. With the notification of successful authentication, the authentication system provides this PMK to the authenticator, the AP, in the MPPE-Key attribute for RADIUS or its equivalent attribute for other authentication systems. The PMK is used to derive a pairwise transient key (PTK) through an exchange of IEEE 802.1X EAPOL-Key frames, often called the *4-Way Handshake* in the IEEE 802.11 standard. The EAPOL-Key frames allow the STA and AP (i.e., the Supplicant and Authenticator) to each supply a *nonce*, i.e., a random value, to the other that is then combined with the PMK, the addresses of the STA and AP, and a fixed string to produce the PTK. Subfields of the PTK are used as the key encryption key (KEK), key confirmation key (KCK), and the temporal key (TK). The KEK and KCK are used during the IEEE 802.1X EAPOL-Key frame exchange to provide confidentiality and authenticity in the final two frames of the 4-Way Handshake. The TK is used as the frame encryption key for protected data frames and is an input to the particular pairwise cipher suite negotiated during the association phase. See Figure 4–4.

A further exchange of IEEE 802.1X EAPOL-Key messages, i.e., a two-way handshake, is used to deliver the group temporal key (GTK) to the STA from the AP. In the first IEEE 802.1X EAPOL-Key frame, from the AP to the STA, the GTK is encrypted with the KEK from the 4-Way Handshake. A message integrity check (MIC) value is calculated using the KCK from the 4-Way Handshake. Because the AP may change the GTK at any time, the KEK and KCK from the 4-Way Handshake must be saved for the duration of a STA's association. When the STA verifies the MIC in the first IEEE 802.1X EAPOL-Key frame, it replies with the last IEEE 802.1X EAPOL-Key frame to acknowledge that it has installed the group key. In an IBSS, the key exchange is more complex and requires both STAs to perform an independent two-way EAPOL-Key exchange. One exchange delivers the GTK used by the first STA to the second STA. The other exchange delivers the GTK used by the second STA to the first STA. See Figure 4–5.

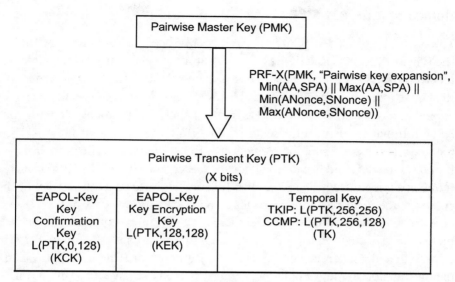

Figure 4–4: Pairwise key hierarchy

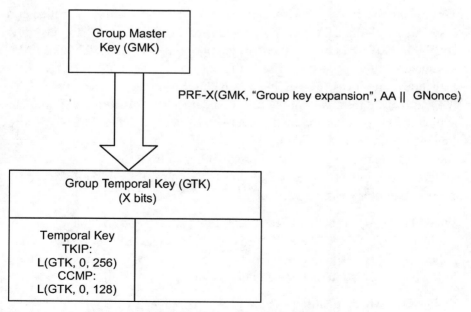

Figure 4–5: Group key hierarchy

Details of IEEE 802.1X EAP AKMP operation

The operation of the IEEE 802.1X EAP AKMP begins long before the first IEEE 802.1X frame is sent. The STA must store the RSN information element received from the AP in the Probe Response or Beacon frame, if no Probe Response frame has been received. This RSN information element must be copied into one of the IEEE 802.1X frames later in the protocol. Copying the RSN information element to an later IEEE 802.1X frame allows the protection of this information element in a frame where modification of the frame can be detected because neither the Beacon nor the Probe Response frame is authenticated. The STA must first authenticate using the legacy open system authentication algorithm. This step must be followed by an Association Request frame that includes an RSN information element that includes a proper subset of the cipher suites and AKM suites from the RSN information element received in the Probe Response or Beacon frame. When the information in the Association Request frame is acceptable to the AP, it will respond with an Association Response frame with a successful status code.

Throughout the process of association, the AP has blocked the controlled port associated with the STA. This port remains blocked and the AP accepts only IEEE 802.1X frames on the uncontrolled port until the IEEE 802.1X AKMP has successfully completed. The STA may begin the IEEE 802.1X authentication process by sending an EAPOL-Start frame to the AP. Alternatively, it may wait for the AP to send it an EAPOL-Request frame. The authentication process continues with the STA sending an EAPOL-Response frame for each received EAPOL-Request frame from the AP. At the end of the authentication process, the AP will send either an EAP-Success or EAP-Failure frame to the STA to indicate the success or failure of the authentication process, respectively.

When the AKMP has completed successfully, the AP and STA have established a PMKSA. The PMKSA is, as described in IEEE 802.11i, "a set of policies and keys used to protect information." The PMKSA consists of the following information that is maintained by both the AP and STA:

- PMKID, which identifies the security association
- Authenticator MAC address

- PMK
- Lifetime
- AKMP
- All authorization parameters specified by the AS or local configuration, e.g., the STA's authorized SSID

EAPOL-Key frames

The exchange of keys between the STA and AP occur in IEEE 802.1X EAPOL-Key frames. The format of the EAPOL-Key frame is shown in Figure 4–6. The EAPOL-Key frame is carried in a IEEE 802.11 data frame using the encapsulation described in 7.3 of the IEEE 802.11 standard. This format supersedes the pre-standard format described in RFC 3580.

SNAP Encapsulation – 8 octets
Descriptor Type – 1 octet
Key Information – 2 octets
Key Length – 2 octets
Key Replay Counter – 8 octets
Key Nonce – 32 octets
EAPOL-Key IV – 16 octets
Key RSC – 8 octets
Reserved - 8 octets
Key MIC – 16 octets
Key Data Length – 2 octets
Key Data – *n* octets

Figure 4–6:   EAPOL-Key frame, key descriptor format

The value of the Descriptor Type field is 2 and indicates that the key descriptor is an IEEE 802.11 key descriptor.

The Key Information field is bit significant and shown in Figure 4–7.

B0	B2 B3 B4	B5	B6	B7	B8	B9	B10	B11	B12 B13	B15
Key Descriptor Version	Key Type	Reserved	Install	Key Ack	Key MIC	Secure	Error	Request	Encrypted Key Data	Reserved

Figure 4–7: Key Information field

The Key Length field is the length, in octets, of the PTK or GTK described by this key descriptor. Valid values for this field are 5 for 40-bit WEP, 13 for 104-bit WEP, 32 for TKIP, and 16 for CCMP.

The value of the Key Replay Counter field is random and chosen to be the first sequence number used by a frame encrypted with the key described in this descriptor. The sequence number is incremented for each subsequent frame encrypted with this same key. The sequence counter is incremented, modulo 2^{64}, until it approaches the value of the Key Replay Counter field. A new key must be established if the sequence number exhausts the sequence number space.

The Key Nonce field is used to exchange the nonce values between the STA and AP to allow them to calculate the PTK. Two nonces are used: one from the STA (SNONCE) and one from the AP (ANONCE). The use of two nonces proves that both devices are "live" and not replaying a frame captured from some earlier exchange between the corresponding STA and AP. The use of two nonces, as long as the use of a nonce is not repeated for a given PMK and corresponding pair, prevents an attacker that is replaying a previous frame from calculating the correct PTK.

The EAPOL-Key IV field is zero, except when a key is present in the Key Data field. When a key is present in the Key Data field, the EAPOL-Key IV field holds the value of the IV used with the KEK to encrypt the Key Data field.

The Key RSC field contains the receive sequence counter value associated with the key described by this key descriptor. This value is used by the AP to establish the initial value of the replay counter for this key in the STA. This field is zero, except in Frame 3 of the 4-Way Handshake and Frame 1 of the two-way Group Key Handshake.

The Key MIC field contains the value of the integrity check function, calculated using the KCK. The key MIC is calculated using the HMAC-Message Digest 5 (MD5) if the Descriptor Type field value is 1 or using the HMAC-SHA-1-128 if the Descriptor Type field value is 2.

The value of the Key Data Length is the length, in bytes, of the Key Data field. The Key Data field is optional. If the Key Data field is not present in the key descriptor, the value of the Key Data Length field is zero.

The Key Data field, if present, contains one or more information elements providing data necessary for establishing the PTK or GTK. For RSN, this field contains one or more RSN information elements. It may contain one or more Vendor-Specific information elements (i.e., element ID 0xDD) for vendor-specific key derivations.

PSK AKMP

PSK AKMP is designed for networks where an authentication system, such as a RADIUS server, is not available. This situation is typically used in small offices and consumer applications. PSK AKMP shortcuts the EAP authentication exchange and uses a secret shared by the AP and STA as the PMK. For best security, the shared secret should be 256 bits and random. However, because this secret must be entered manually into all APs and STAs in the WLAN, an informative annex in the IEEE 802.11 standard describes an algorithm that can be used to transform a pass phrase into a 256-bit string. The STA and AP utilize this calculated 256-bit string and the IEEE 802.1X EAPOL-Key 4-Way Handshake to establish the PTK and GTK. The pairwise and group key hierarchies are identical.

While PSK is much more convenient and potentially much less expensive for small WLANs than installing a separate authentication system (e.g., an AP may include an embedded authentication system), it is also more prone to compromising its security. As much care must be taken to maintain the security of the shared secret as is necessary with WEP shared keys. PSK is also subject to offline dictionary attacks against the shared secret.

If the shared secret becomes known to an attacker, the security of the entire network is compromised. With the shared secret, an attacker can successfully authenticate with the WLAN. An attacker can also eavesdrop on the IEEE 802.1X EAPOL-Key exchange between the AP and other STAs and calculate the PTK they will use. This knowledge allows the attacker to decrypt all traffic without exposing itself. If individual users have knowledge of the shared secret, changing that shared secret using PSK on all the equipment in the WLAN whenever someone departs a company is the only way to ensure that the WLAN remains secure. This necessity adds a significant burden to the network administration staff of an enterprise. The consequences of PSK use should be very carefully considered before selecting it to protect a network.

Details of PSK AKMP operation

There is no explicit authentication or verification of identity when using PSK because the PSK is shared by many STAs. The authentication of a STA and AP is implicit and assumed upon successful completion of the 4-Way Handshake to establish a PTK between them. What the IEEE 802.1X exchange demonstrates is that both the STA and AP have the same PSK. The exchange is begun by the STA sending an EAPOL-Start frame to the AP. The AP responds with an EAP-Success frame, followed by the first EAPOL-Key frame. The 4-Way Handshake proceeds as already described for IEEE 802.1X authentication above. The PSK value is used to create the PMK in the key hierarchy and generate the KEK, KCK, and TK. A further two-way Group Key Handshake is used by the AP to provide the GTK to the STA.

IEEE 802.11i has also defined a function to generate the PMK from the PSK in order to try to get as much randomness out of what is usually a password that is too short to provide any real security. This function is only informative; but if a product allows entry of a simple ASCII string for the PSK, it is highly recommended that the function be used to transform this string into a bit string that is used as the PMK. It does not add any more randomness (or entropy) than is present in the original ASCII string, but it will at least distribute that randomness throughout the entire length of the PMK. Use of this function is required by the Wi-Fi Alliance to obtain certification of an IEEE 802.11i product. Because the PSK is vulnerable to offline dictionary attacks, it is highly recommended that a minimum length of 20 characters be required for strong PSKs. There are already publicly available attack tools against the PSK that require only the frames captured from a single EAPOL-Key exchange.

PMK caching

Because the time required to complete a full EAP authentication exchange can be long, IEEE 802.11i includes a method to remember the PMKSAs that have been developed between an AP and STA. PMK caching requires an AP to keep the PMKSAs for STAs that have been authenticated or preauthenticated, although IEEE 802.11i does not place any requirement on how many of these PMKSAs must be kept in the cache nor how long they must be kept in the cache. A STA must also keep a cache of PMKSAs that it has developed with APs. Only if both the AP and STA have matching PMKSAs will PMK caching allow the EAP authentication exchange to be bypassed and the AP and STA to proceed directly to the key exchange. A STA indicates to an AP that is has one or more cached PMKSAs by including a list of PMKIDs in its Reassociation Request frame to the AP. The AP examines the PMKSAs it has cached for the reassociating STA and determines whether any PMKIDs match the associations identified by the PMKIDs in the STA's Reassociation Request frame. If there is a matching PMKSA, the AP sends EAPOL-Key Message 1 to indicate to the STA the PMKSA to use. If there is no matching PMKSA, the AP sends an EAP Identity Request frame to the STA to indicate the requirement to complete the entire EAP authentication exchange.

Preauthentication

When using either IEEE 802.1X or PSK AKMPs, IEEE 802.11i provides a means for a STA to preauthenticate with APs other than the one with which it is currently associated. The STA must first be fully authenticated with its current AP, including any and all key exchanges that are required. Once the STA is authenticated, it may use the IEEE 802.1X or PSK AKMP to preauthenticate with other APs of which the STA has knowledge. The STA typically receives a Beacon or Probe Response frame from the other APs in order to have a copy of the RSN information element to include in the appropriate message of the AKMP. The IEEE 802.11 Task Group k (TGk) is developing a mechanism to provide a "neighbor report" to a STA. This report will indicate information about neighboring APs. This information would not be authoritative because it would not be authenticated. However, since the RSN information element is validated in the EAPOL-Key exchange, this information could be used in lieu of receiving the RSN information element in a Beacon or Probe Response frame.

When the STA has the information required for preauthentication, it can begin the appropriate IEEE 802.1X or PSK authentication exchange by using the current AP at which it is associated as a conduit to deliver the frames to the other AP. The frames used for preauthentication are identical to the frames used when authenticating directly with an AP, except that the Ethertype value is changed to 88-c7. This new value is used to allow Ethernet switches to implement a policy of not forwarding frames with the original IEEE 802.1X Ethertype to prevent spoofing. At the end of the preauthentication, the STA and AP with which it has preauthenticated have established a PMKSA and PMKID. The PMKSA and PMKID are cached at the AP after completing the preauthentication. The STA can simultaneously exchange data through its currently associated AP to any network device beyond the AP as it is preauthenticating with one or more additional AP.

The benefit of preauthentication is that the STA can reduce the time required before it may transfer data through an AP with which it is reassociating, by removing the need to complete a full authentication from the beginning. When reassociating, the STA presents its PMKID; and if the AP has a PMKSA that matches the PMKID presented by the STA, the AP proceeds directly to the key exchange handshake. If the key exchange is completed successfully, the STA may immediately begin exchanging data frames protected by the negotiated cipher suites. If the key exchange does not complete successfully, the AP will delete the cached PMKSA and PMKID; and the STA must complete a full authentication.

TRANSITION SECURITY NETWORKS (TSNs)

TSNs provide a method to use legacy equipment that is capable only of WEP encryption and to use equipment that provides RSN capabilities in a mixed environment. This capability allows equipment to be upgraded or replaced gradually when moving an existing WLAN to IEEE 802.11i and is the only difference between an RSN and a TSN. A TSN still advertises RSN capabilities, authentication suite, and cipher suites in an RSN information element in its Beacon and Probe Response frames. However, instead of responding with an association failure status code upon receipt of an Association Request frame without an RSN information element, an AP in a TSN can accept the Association Request frame. An AP configured for a TSN can accept an association request from a non-RSN STA only if there is an appropriate WEP key configured for the associating STA or if IEEE 802.1X is configured as the AKMP. An appropriate WEP key can be either a shared key or a key mapping key for the MAC address of the STA. If an appropriate WEP key is not configured for the STA requesting association and IEEE 802.1X is not configured as the AKMP, the Association Response frame will contain a status code indicating the association failed due to failed security policy (status code 46).

CONFIDENTIALITY: NEW ENCRYPTION ALGORITHMS

IEEE 802.11i describes two new confidentiality algorithms. The first algorithm is specifically designed to enable equipment manufactured prior to the availability of IEEE 802.11i to be upgraded with significantly stranger security. This first algorithm is named *Temporal Key Integrity Protocol* (TKIP). The second algorithm is designed to take advantage of the recently adopted advanced encryption system (AEP), providing security sufficient to protect information carried by IEEE 802.11 WLANs for the foreseeable future.

FIXING WEP: TEMPORAL KEY INTEGRITY PROTOCOL (TKIP)

Fixing WEP is a difficult thing to do. It comes with some significant constraints. The goal of TGi was to enable manufacturers to upgrade already fielded devices to the new, more secure version of WEP through a firmware or driver update, while eliminating the known vulnerabilities and minimizing the risk that any new vulnerability would be found in the upgrade. Manufacturers and cryptographers cooperated to identify the requirements for backward compatibility with existing IEEE 802.11 hardware. These backward compatibility requirements presented the most significant hurdles. In particular, the constraints to continue using the RC4 cipher, for which many implementations have hardware acceleration, and to require minimal additional processing, because several implementations run on embedded processors with little spare processing power, required a number of tradeoffs and compromises.

TGi developed the TKIP specification to meet these requirements. TKIP provides significantly improved privacy and authentication, while still using the RC4 cipher and requiring little additional processing power. It has been fielded by most manufacturers and shown to run on their already deployed hardware, although the wave of upgrades to TKIP did not materialize for several reasons. The expected upgrades to legacy equipment did not happen at least because of some of the compromises required by TKIP. To meet the processor requirements, a relatively weak MIC was included in TKIP. The TKIP MIC is subject to attacks that would allow modification of a frame without detection in a reasonably short time. To reduce the effect of this attack, TKIP requires countermeasures to be invoked when more than one MIC failure is detected within 1 min. These countermeasures are relatively onerous; they require that the WLAN be disabled for a minute and that all STAs be disassociated. These requirements led a number of manufacturers to develop a proprietary alternative to TKIP, called *CKIP*, that uses a different MIC that is not subject to the same degree of attack. Nonetheless, TSN has been adopted by the Wi-Fi Alliance as a requirement to obtain Wi-Fi® interoperability certification. Wi-Fi has named TKIP to be Wi-Fi Protected Access (WPA). WPA is drawn from an early draft of IEEE 802.11i. There are several significant differences between WPA and the ultimately adopted TKIP of IEEE 802.11i, including differences in the information elements and EAPOL-Key format.

TKIP is a combination of three items: the RC4 stream cipher, the Michael MIC function, and rapid key rotation. The Michael MIC addresses the simple message injection attack against WEP. Rapid key rotation addresses the weakness of static, long-lived keys of WEP. These two new features address the published attacks against the original WEP algorithm.

TKIP operation

TKIP operation begins with the MSDU from the layer above the MAC. The MSDU is concatenated with the DA, SA, and requested priority. The Michael MIC is then calculated over this concatenation and appended to it. The resulting extended MSDU, i.e., the original MSDU plus the 8 octets of MIC, is then handled by the IEEE 802.11 MAC as any other MSDU. This process includes fragmentation, if required. The MIC is calculated over the original MSDU to prevent an attack where one MPDU of a stream of fragments could be substituted without detection. This approach also allows flexibility in the implementation of the MIC and allows it to be placed either in a host processor driver or in the network interface card as firmware or hardware. The result is shown in Figure 4–8.

When transmitting a frame using the TKIP cipher, the transmitting MAC must determine the value of the key to use to initialize the RC4 stream cipher. TKIP calculates this key in two phases. Both phases make use of a transmit sequence counter (TSC) that is 48 bits in length and initialized to a value of one when a new TK is established. The TSC is a monotonically increasing counter. Each new frame or fragment to be transmitted is assigned the next unused value from the TSC and encrypted with a key determined from this value.

The first phase uses the 32 MSBs of the TSC and combines it with the TA and TK to obtain the TTAK. The result of the TTAK calculation is an 80-bit string. The TTAK can be cached and used repeatedly because the upper portion of the TSC changes only once for each 2^{16} frames sent from the TA.

The second phase to determine the key necessary to encrypt a frame for transmission is to calculate the WEP seed. The WEP seed is used to initialize the RC4 algorithm. The inputs to the WEP seed calculation are the TK, TTAK, and TSC. Only the 16 LSBs of the TSC are used to calculate the WEP seed. The result of the WEP seed calculation is a 128-bit string. The WEP seed can be precalculated because it is dependent only on the TTAK, TK, and TSC.

A particular value of WEP seed is used to encrypt only a single frame and then is discarded. The strength of the TKIP algorithm relies on the usage of a particular RC4 encryption key (i.e., WEP seed) only once. Reuse of encryption keys would significantly weaken the security of TKIP.

6	6	1	3	M	1	1	1	1	1	1	1	1	octets
DA	SA	Priority	0	Data	M0	M1	M2	M3	M4	M5	M6	M7	

Figure 4–8: TKIP MIC processing format

Michael MIC

The Michael MIC is a compromise between strength and computational complexity. Once again, because of the limited processing power available in some legacy implementations, the MIC is designed to provide a degree of protection against modification of the encrypted information that is not as strong as is really desired. The MIC is calculated over the MSDU, before the MSDU might be fragmented for transmission. To protect information in the MAC header, the DA, SA, and Priority field (defined by IEEE P802.11e) from the header are also included in the MIC calculation. The MIC is 8 bytes long and increases the length of the MSDU by 8 bytes. This added length must be taken into account when determining whether a frame must be fragmented. To avoid fragmenting frames that are encrypted with TKIP, the fragmentation threshold should be set to a value of 2363 or larger.

TKIP-encrypted frame description

The format of a frame encrypted using TKIP varies somewhat from the original format for a WEP-encrypted frame. The definition of the original IV field of the WEP-encrypted frame is modified, and an optional Extended IV field may be present. The format of the new TKIP-encrypted frame is shown in Figure 4–9.

The IV/KeyID field is 4 octets long. It comprises several subfields. The first octet is the second least significant byte of the TSC (i.e., bits 8–15 of the TSC, or TSC1). The second octet is named WEP Seed[1]. It is derived from TSC1 by the following formula:

WEP Seed[1] = (TSC1 | 0x20) & 0x7f; where "|" represents the bitwise logical "or" function and "&" represents the bitwise logical "and" function.

The third octet of the IV/KeyID field is the least significant byte of the TSC (i.e., bits 0–7 of the TSC, or TSC0). The final octet of the IV/KeyID field is itself composed of subfields. Bits 7–6 of this octet are the WEP Key ID bits. These bits function exactly as they do for WEP. Bit 5 is the Extended IV bit. When the Extended IV bit is equal to one, the 4 bytes following the IV/KeyID field contain the 32 MSBs of the TSC.

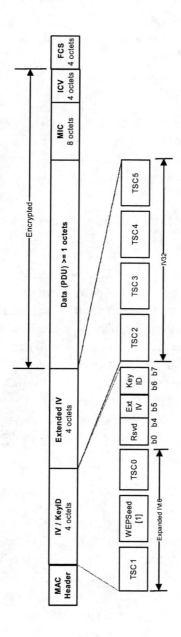

Figure 4–9: TKIP-encrypted frame format

The IV/KeyID field and the Extended IV field are never encrypted. The first octet of a frame that is encrypted is the Frame Body field. The last octet that is encrypted is the WEP ICV field. Between these two fields is the TKIP MIC. The TKIP MIC is encrypted as well.

Attack countermeasures

The Michael MIC is relatively weak. For this reason, IEEE 802.11i defines countermeasures to be taken when an attack against the MIC is suspected. The countermeasures are to deauthenticate all associated STAs and not accept any new associations for 60 s as well as to discard the current GTK if the group cipher is TKIP.

Countermeasures are invoked when two MIC failures are detected within a 60 s window. A MIC failure occurs when an MSDU is reconstructed from successfully received MPDUs without WEP ICV failures, yet the MIC calculated for the MSDU and the MIC received with the MDSU do not match. This event is extraordinarily unlikely to occur by chance because the changes to the MPDUs would have to result in both the ICV and FCS of the frames being correct. Having the same event happen twice in a 60 s window is nearly impossible. MIC failures can be detected locally or at a peer MAC. When detected at a peer MAC, MIC failures are reported in an EAPOL-Key frame with the Request, Error, and MIC bits set in the Key Information field. A STA sends an EAPOL-Key frame to the AP with these bits set to indicate the MIC failure that was detected and to request an immediate rekeying. The EAPOL-Key frame must be encrypted with the current PTK as would any other frame sent to the AP at that time.

NEW TRANSFORMS

IEEE 802.11i defines the use of a new encryption transform to protect the confidentiality, authenticity, and integrity of IEEE 802.11 frames. It also provides protection from replay attacks. The new transform is the advanced encryption system (AEP) using Counter-Mode with Cipher Block Chaining (CBC) Message Authentication Code (MAC) Protocol. This concept is abbreviated to AES CCMP or often just CCMP. Confidentiality, authenticity, message integrity, and replay protection are achieved in a single transform.

AES is in a different class of cipher algorithms from RC4. Where RC4 is a stream cipher, able to generate a continuous, pseudo-random sequence of bytes to be used to encrypt a plaintext of arbitrary length, AES is a block cipher able to transform a fixed-length block of plaintext into an identical length of ciphertext. In CCMP, the block length used by the algorithm is 128 bits. The key length is also 128 bits.

RFC 3610 describes the details of the operation of counter-mode with CBC MAC. IEEE 802.11i refers to that RFC for the detailed description of the algorithm. The RFC provides a description of the algorithm that requires two parameters to complete the description for a particular application, such as IEEE 802.11i CCMP. The first parameter is the length of the MAC (i.e., the MIC in CCMP, since MAC is used for another purpose by IEEE 802.11). For CCMP, the value of this parameter is 8, the number of octets in the MIC. The second parameter is the size of the Message Length field. For CCMP, the value of this parameter is 2 and indicates a field that is 2 bytes long. This parameter is sufficient to indicate the length, in bytes, for any legal IEEE 802.11 frame.

RFC 3610 allows the inclusion of information that is not encrypted, but is authenticated, such as header information that is sent as plaintext. CCMP utilizes this feature to include the following information as additional authenticated data (AAD).

- Frame Control field with the following subfields forced to zero: Subtype, Retry, Power Management, and More Data. The Protected Frame bit (formerly WEP bit) is forced to one.
- Address 1, Address 2, and Address 3.
- Sequence Control field with the sequence number forced to zero.
- Address 4.
- QoS Control field, currently reserved and forced to zero. This field is filled by the value defined by IEEE P802.11e.

The Address 4 and QoS Control fields are included in the AAD only when they are present in the MAC header. Without these two fields, the AAD is 22 octets long. The Address 4 field adds 6 bytes when present. The QoS Control field adds 2 bytes when present.

It is important to realize that CCMP does not repeat the information in the AAD elsewhere in the frame. It includes this information only in internal calculations of the MIC. Implementations of CCMP must provide a means, in hardware or software, to feed this information into the authentication calculation. The implementation must be flexible enough to recognize when optional fields are present in the MAC header and to adapt the MIC calculation to include these optional fields as necessary.

CCMP-encrypted frame description

The format of a frame encrypted using CCMP varies slightly from the original WEP-encrypted frame. But it is very similar to the format of the TKIP-encrypted frame, including modifications to the IV field and addition of an Extended IV field. The format of the CCMP-encrypted frame is shown in Figure 4–10. CCMP increase the length of a data frame by 16 bytes by including an additional header and the MIC. To avoid fragmenting a CCMP-encrypted frame, the fragmentation threshold should be set to a value of 2363 or larger.

The CCMP Header field is a fixed length of 8 bytes. It comprises the following subfields: 48-bit Packet Number (PN), 2-bit KeyID, and 1-bit Extended IV indicator. It immediately follows the MAC header and is not encrypted.

The PN is used as part of the nonce during CCMP processing of a frame. It is a 48-bit, monotonically increasing counter. It is functionally identical to the TSC used by TKIP. Each MPDU must use a fresh value from this counter, and no values from this counter must ever be reused with the same TK. Failure to heed these requirements severely reduces the security of the CCMP algorithm. The PN is divided in 6 octets, PN0 through PN5, where PN0 is the least significant octet and PN5 is the most significant octet. PN0 and PN1 are placed in the first 2 bytes of the CCMP Header field. PN2 through PN5 are placed in the last 4 bytes of the CCMP Header field.

The reserved byte is set to zero.

The fourth byte of the CCMP Header field comprises three subfields. Bits 0–4 are reserved and set to zero. Bit 5 is the Extended IV bit. It is set to 1 for all

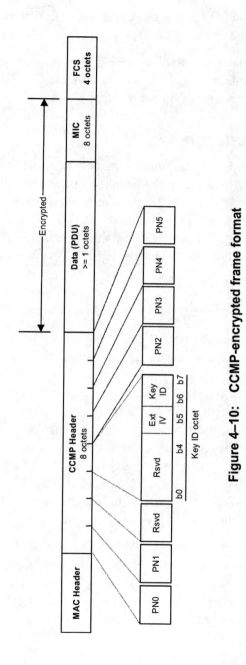

Figure 4–10: CCMP-encrypted frame format

CCMP frames. This setting is to ease implementations that provide both TKIP and CCMP. Bits 6 and 7 are the KeyID bits.

Following the CCMP Header field is the encrypted frame body and MIC. The frame body must include at least 1 byte in addition to the MIC. In other words, a null data frame is never encrypted.

SECURITY MANAGEMENT

With the enhancements to authentication and confidentiality in IEEE 802.11i come many new management attributes in the MIB. These attributes are necessary to allow a remote management entity to determine the configuration of a STA implementing IEEE 802.11i and to monitor the operation of the IEEE 802.11i-related functions of the STA. This section will describe the changes and additions to the IEEE 802.11 MIB that are made by IEEE 802.11i.

Changes to existing attributes and tables

Two significant changes are made by IEEE 802.11i to the interpretation of existing MIB attributes. The first change is that the dot11PrivacyInvoked attribute now is used to indicate whether any confidentiality algorithm is used to protect data frames. When this attribute is true, data frames are protected by a confidentiality algorithm. When this attribute is false, data frames are not protected. The previous meaning of this attribute, that WEP was used to protect data frames when the value of the attribute is true, must now be determined by testing for the presence and value of a new attribute, dot11RSNAEnabled. Only if dot11RSNAEnabled is false or is not present, can dot11PrivacyInvoked be interpreted to indicate that WEP is the confidentiality algorithm used to protect data frames.

The second change is that the dot11WEPICVErrorCount attribute counts only ICV errors for frames protected by the original WEP confidentiality algorithm. Even though TKIP utilizes WEP as the underlying algorithm for encrypting an individual frame, this attribute will not count ICV errors for frames encrypted using TKIP.

IEEE 802.11i adds two attributes to existing attribute groups. The dot11RSNAOptionImplemented attribute is added to the Dot11StationConfigEntry attribute group. This new attribute is a truth value, indicating that the STA has implemented IEEE 802.11i RSNA and is capable of RSNA operation. The second new attribute is dot11RSNAEnabled. This attribute is another truth value and is added to the Dot11PrivacyEntry attribute group. When the dot11RSNAEnabled attribute is true, the STA will send the RSN information element in Beacon and Probe Response frames. When this attribute is true, the dot11PrivacyInvoked attribute must also be true.

New attributes and tables

IEEE 802.11i adds several new tables of attributes for the management of the new security functionality. The tables are used for configuration of security parameters, lists of pairwise cipher suites, lists of AKM suites, and a table of statistics.

Dot11RSNAConfigTable

The Dot11RSNAConfigTable holds the attributes pertaining to the configuration of the operation of RSNA functions of the STA. The table consists of a row of attributes for each IEEE 802.11 interface on the STA. The attributes in the row are described in this section.

The dot11RSNAConfigVersion attribute is a 32-bit integer. The attribute is read-only. Its value is the highest revision of RSNA that the STA supports. The current version is 1. All other values are reserved.

The dot11RSNAConfigPairwiseKeysSupported attribute is a 32-bit integer. The attribute is read-only. Its value indicates the number of simultaneous pairwise keys the STA supports for RSNA.

The dot11RSNAConfigGroupCipher attribute is a 4-byte octet string. This attribute is read-only. It is used to control the group cipher suite used by the STA. In an AP or a STA starting an IBSS, the value of this attribute will be copied into the Group Cipher field of the RSN information element. The format of this attribute value is a 3-byte OUI followed by a 1-byte suite selector. The values for the OUI and suite selector are identical to the values in the definition of the field in the RSN information element.

The dot11RSNAConfigGroupRekeyMethod attribute is an enumerated integer. The attribute is read-write. The value of the attribute indicates the group cipher rekey method to be used. The allowable values indicate rekeying is disabled (1), rekeying is entirely based on elapsed time (2), rekeying is entirely based on the number of packets transmitted under the key (3), and rekeying is based on either elapsed time or number of packets transmitted under the key (4). Rekeying based on elapsed time, i.e., either method 2 or 4, causes the group cipher to be rekeyed once per day. Rekeying based on number of packets transmitted, i.e., either method 3 or 4, causes the group cipher to be rekeyed before the sequence number space is exhausted. Rekeying is applicable only to STAs that performed the authenticator role in the IEEE 802.1X authentication exchange.

The dot11RSNAConfigGroupRekeyTime attribute is an unsigned 32-bit integer. The attribute is read-write. The value of the attribute indicates the number of seconds that may elapse before the group key cipher must be rekeyed. The default value for this attribute is 86 400 s, or one day.

The dot11RSNAConfigRekeyPackets attribute is an unsigned 32-bit integer. This attribute is read-write. The value of the attribute indicates the number of packets that may be transmitted using the group key before it must be rekeyed. The attribute value is multiplied by 1000 to obtain the number of allowable packets.

The dot11RSNAConfigRekeyStrict attribute is a truth value. This attribute is read-write. When this attribute is true, the group key cipher shall be rekeyed whenever a STA leaves the BSS. Such departure can be a difficult thing to determine because a STA can leave a BSS without notifying the AP. The description of this attribute in IEEE 802.11i seems to indicate that the AP must rekey the group cipher whenever it determines that a STA has left the BSS either by receipt of a Disassociation frame or by any internal method of determination.

The dot11RSNAConfigPSKValue attribute is a 32-byte octet string. This attribute is defined as read-write, but is actually write-only according to the description in IEEE 802.11i. This attribute holds the value of the PSK. This value may be entered directly or indirectly through the PSK pass-phrase in the dot11RSNAConfigPSKPassPhrase attribute.

The dot11RSNAConfigPSKPassPhrase attribute is a display string. The attribute is defined as read-write, but is actually write-only according to the description in IEEE 802.11i. This attribute holds the value of the PSK pass-phrase. The pass-phrase is processed as described in the annex of IEEE 802.11i to result in the actual PSK in the dot11RSNAConfigPSKValue attribute.

The dot11RSNAConfigTSNEnabled attribute is a truth value. The attribute is read-write. When dot11PrivacyInvoked, dot11RSNAEnabled, and dot11RSNAConfigTSNEnabled attributes are true, associations by pre-RSNA STAs are allowed. When the value of the dot11RSNAConfigTSNEnabled attribute is false and both dot11PrivacyInvoked and dot11RSNAEnabled are true, associations by pre-RSNA STAs are not allowed.

The dot11RSNAConfigGroupMasterRekeyTime attribute is an unsigned 32-bit integer. This attribute is read-write. The value of this attribute indicates the elapsed time, in seconds, after which the group master key must be changed. The default value for this attribute is 604 800 s, or 7 days.

The dot11RSNAConfigGroupUpdateTimeOut attribute is an unsigned 32-bit integer. This attribute is read-write. The value of the attribute indicates the time to elapse, in milliseconds, before Message 1 of the group key message exchange may be retransmitted. The default value of this attribute is 100 ms.

The dot11RSNAConfigGroupUpdateCount attribute is an unsigned 32-bit integer. This attribute is read-write. The value of this attribute indicates the number of times that Message 1 of the group key message exchange will be retransmitted to an individual STA before it is abandoned. The total number of attempts to send Message 1 of the group key message exchange to a STA is one more than the value of this attribute. The default value of this attribute is 3.

The dot11RSNAConfigPairwiseUpdateTimeOut attribute is an unsigned 32-bit integer. The attribute is read-write. The value of this attribute indicates the elapsed time, in milliseconds, after the transmission of Message 1 or

Message 3 of the 4-Way Handshake before the message will be retransmitted. The default value of this attribute is 100 ms.

The dot11RSNAConfigPairwiseUpdateCount attribute is an unsigned 32-bit integer. This attribute is read-write. The value of this attribute indicates the number of times that Message 1 or Message 3 of the 4-Way Handshake will be retransmitted to an individual STA before it is abandoned. The total number of attempts to send each of these messages to a STA is one more than the value of this attribute. The default value of this attribute is 3.

The dot11RSNAConfigGroupCipherSize attribute is an unsigned 32-bit integer. This attribute is read-only. The value of this attribute indicates the length, in bits, of the group cipher key.

The dot11RSNAConfigPMKLifetime attribute is an unsigned 32-bit integer. This attribute is read-write. The value of this attribute indicates the maximum lifetime, in seconds, of a PMK in the PMK cache. The default value of this attribute is 43 200 s, or 12 h.

The dot11RSNAConfigPMKReauthThreshold attribute is an unsigned 32-bit integer. This attribute is read-write. The allowable range of values for this attribute is from 1 to 100. The value of this attribute is the percentage of the time of the dot11RSNAConfigPMKLifetime attribute that may elapse before an IEEE 802.1X reauthentication is begun. The default value of this attribute is 70%.

The dot11RSNAConfigNumberofReplayCounters attribute is an unsigned 32-bit integer. This attribute is read-only. While this attribute is not an enumerated attribute, the description in IEEE 802.11i does define only four allowable values. The allowable values for this attribute define the number of replay counters available for each security association. The number of replay counters is one (0), two (1), four (2), or sixteen (3).

The dot11RSNAConfigSATimeout attribute is an unsigned 32-bit integer. This attribute is read-write. The value of this attribute indicates the time, in seconds, during which the setup of a security association must be completed before it is abandoned and must restart from the beginning. The default value of this attribute is 60 s.

The dot11RSNAConfigAuthenticationSuiteSelected attribute is a 4-byte octet string. This attribute is read-only. The value of this attribute indicates the authentication suite negotiated during the association exchange. The format of this attribute is identical to the authentication suite selectors in the RSN information element. While this attribute is useful for a single STA, it is not clear what value should be indicated here by an AP that advertises more than one authentication suite.

The dot11RSNAConfigPairwiseCipheSuiteSelected attribute is a 4-byte octet string. This attribute is read-only. The value of this attribute indicates the pairwise cipher suite negotiated during the association exchange. The format of this attribute is identical to the cipher suite selectors in the RSN information element. While this attribute is useful for a single STA, it is not clear what value should be indicated here by an AP that advertises more than one pairwise cipher suite.

The dot11RSNAGroupCipheSuiteSelected attribute is a 4-byte octet string. This attribute is read-only. The value of this attribute indicates the group cipher suite negotiated during the association exchange. The format of this attribute is identical to the cipher suite selectors in the RSN information element. While this attribute is useful for a single STA, it is not clear what value should be indicated here by an AP that advertises more than one group cipher suite.

The dot11RSNAPMKIDUsed attribute is a 16-byte octet string. This attribute is read-only. The value of this attribute indicates the PMKID used in the last 4-Way Handshake.

The dot11RSNAAuthenticationSuiteRequested attribute is a 4-byte octet string. This attribute is read-only. The value of this attribute indicates the selector of the most recent authentication suite requested. The format of this attribute is identical to the authentication suites in the RSN information element.

The dot11RSNAPairwiseCipherRequested attribute is a 4-byte octet string. This attribute is read-only. The value of this attribute indicates the selector of the most recent pairwise cipher suite requested. The format of this attribute is identical to the cipher suites in the RSN information element.

The dot11RSNAGroupCipherRequested attribute is a 4-byte octet string. This attribute is read-only. The value of this attribute indicates the selector of the most recent group cipher suite requested. The format of this attribute is identical to the cipher suites in the RSN information element.

Pairwise Cipher Suites Table

IEEE 802.11i defines the dot11RSNAConfigPairwiseCiphersTable to list the pairwise ciphers available in an implementation. Each row in this table represents an individual cipher suite and its configuration. The individual attributes in each row are described in this section.

The dot11RSNAConfigPairwiseCipher attribute is a 4-byte octet string. The attribute is read-only. The value of this attribute indicates the selector of the pairwise cipher. The format of this attribute is identical to the cipher suites in the RSN information element.

The dot11RSNAConfigPairwiseCipherEnabled attribute is a truth value. The attribute is read-write. When the value of this attribute is true, the pairwise cipher indicated by the value of the dot11RSNAConfigPairwiseCipher attribute will appear in the list of pairwise ciphers in the RSN information element in Beacon and Probe Response frames from an AP or STA starting an IBSS. A STA joining a BSS shall send in its Association Request frames only pairwise cipher suites for which the cipher is enabled by the dot11RSNA-ConfigPairwiseCipherEnabled attribute. When the value of this attribute is false, the corresponding pairwise cipher does not appear in the RSN information element.

The dot11RSNAConfigPairwiseCipherSize attribute is an unsigned 32-bit integer. The attribute is read-only. The value of this attribute indicates the key size to be used with the corresponding pairwise cipher. While this attribute is not enumerated, two values are defined: 256 for TKIP and 128 for CCMP.

AKM Suites Table

Similarly to the pairwise cipher suites, IEEE 802.11i defines the dot11RSNAConfigAuthenticationSuitesTable to list the available AKM suites. Each row in the table represents an individual authentication suite and its configuration. The individual attributes are described in this section.

The dot11RSNAConfigAuthenticationSuite attribute is a 4-byte octet string. The attribute is read-only. The value of this attribute indicates the selector of the authentication suite. The format of this attribute is identical to the authentication suites in the RSN information element.

The dot11RSNAConfigAuthenticationSuiteEnabled attribute is a truth value. This attribute is read-write. When the value of this attribute is true, the authentication suite indicated by the value of the dot11RSNAConfig-AuthenticationSuite attribute will appear in the list of authentication suites in the RSN information element in Beacon and Probe Response frames from an AP or STA starting an IBSS. A STA joining a BSS shall send in its Association Request frames only authentication suites for which the suite is enabled by the dot11RSNAConfigAuthenticationSuiteEnabled attribute. When the value of this attribute is false, the corresponding authentication suite does not appear in the RSN information element.

RSNA Statistics Table

In order to monitor the operation of the new IEEE 802.11i security functions, the dot11RSNAStatsTable contains a number of attributes to expose the information required. The table is organized in rows with one row for each associated STA. The attributes in each row are described below.

The dot11RSNAStatsSTAAdress attribute is a 6-byte MAC address. The attribute is read-only. The value of this attribute is the MAC address of the STA to which the rest of the statistics in this row belong.

The dot11RSNAStatsVersion attribute is an unsigned 32-bit integer. The attribute is read-only. The value of this attribute indicates the RSNA version of the corresponding STA as indicated in the RSN information element of its Association Request frame.

The dot11RSNAStatsSelectedPairwiseCipher attribute is a 4-byte octet string. This attribute is read-only. The value of this attribute is a cipher suite selector and indicates the pairwise cipher suite selected by the corresponding STA as indicated in the RSN information element of its Association Request frame. The format of this attribute is identical to the cipher suite selector in the RSN information element.

The dot11RSNAStatsTKIPICVErrors attribute is a 32-bit counter. The attribute is read-only. The value of this attribute indicates the number of ICV errors detected for the corresponding STA if the dot11RSNAStatsSelected-PairwiseCipher attribute indicates the selected cipher suite is TKIP. The value of the dot11RSNAStatsTKIPICVErrors attribute is undefined if the selected cipher site is not TKIP.

The dot11RSNAStatsTKIPLocalMICFailures attribute is a 32-bit counter. The attribute is read-only. The value of this attribute indicates the number of TKIP MIC failures detected in frames received by the corresponding STA if the dot11RSNAStatsSelectedPairwiseCipher attribute indicates the selected cipher suite is TKIP. The value of the dot11RSNAStatsTKIPLocalMIC-Failures attribute is undefined if the selected cipher suite is not TKIP.

The dot11RSNAStatsTKIPRemoteMICFailures attribute is a 32-bit counter. This attribute is read-only. The value of this attribute indicates the number of TKIP MIC failures reported to the STA by a correspondent STA if the dot11RSNAStatsSelectedPairwiseCipher attribute indicates the selected cipher suite is TKIP. The value of the dot11RSNAStatsTKIPRemote-MICFailures attribute is undefined if the selected cipher suite is not TKIP.

The dot11RSNAStatsTKIPCountermeasuresInvoked attribute is a 32-bit counter. This attribute is read-only. The value of this attribute indicates the number of times that countermeasures have been invoked, either locally or remotely, due to multiple TKIP MIC failures in a 60 s period if the dot11RSNAStatsSelectedPairwiseCipher attribute indicates the selected cipher suite is TKIP. The value of the dot11RSNAStatsTKIP-CountermeasuresInvoked attribute is undefined if the selected cipher suite is not TKIP.

The dot11RSNAStatsCCMPFormatErrors attribute is a 32-bit counter. The attribute is read-only. The value of this attribute indicates the number of MSDUs received with an invalid format if the dot11RSNAStatsSelected-PairwiseCipher attribute indicates the selected cipher suite is CCMP. The value of the dot11RSNAStatsCCMPFormatErrors attribute is undefined if the selected cipher suite is not CCMP.

The dot11RSNAStatsCCMPReplays attribute is a 32-bit counter. The attribute is read-only. The value of this attribute indicates the number of received unicast fragments discarded as being replayed frames if the dot11RSNAStatsSelectedPairwiseCipher attribute indicates the selected cipher suite is CCMP. The value of the dot11RSNAStatsCCMPReplays attribute is undefined if the selected cipher suite is not CCMP.

The dot11RSNAStatsCCMPDecryptErrors attribute is a 32-bit counter. The attribute is read-only. The value of this attribute indicates the number of received fragments, either unicast or multicast, discarded as being undecryptable if the dot11RSNAStatsSelectedPairwiseCipher attribute indicates the selected cipher suite is CCMP. The value of the dot11RSNA-StatsCCMPDecryptErrors attribute is undefined if the selected cipher suite is not CCMP.

The dot11RSNAStatsTKIPReplays attribute is a 32-bit counter. The attribute is read-only. The value of this attribute indicates the number of received unicast fragments discarded as being replayed frames if the dot11RSNAStats-SelectedPairwiseCipher attribute indicates the selected cipher suite is TKIP. The value of the dot11RSNAStatsTKIPReplays attribute is undefined if the selected cipher suite is not TKIP.

The dot11RSNAStats4WayHandshakeFailures attribute is a 32-bit counter. This attribute is read-only. The value of this attribute indicates the number of times the 4-Way Handshake did not reach a successful conclusion.

Chapter 5   IEEE P802.11e quality of service (QoS) enhancements[1]

BACKGROUND: WHAT IS QOS AND WHY IS QOS NEEDED?

Wireless networks based on IEEE 802.11 have been widely adopted by home users and businesses. New applications such as video and multimedia streaming bring unique and challenging QoS requirements. The ever-expanding demand for bandwidth has caused network congestion and slowdowns, but home users want multimedia distribution to work perfectly without dropouts or glitches. Network administrators need mechanisms to ensure that applications with stringent QoS requirements will function properly over a congested network. These developments have triggered the development of a QoS enhancement for the IEEE 802.11 WLAN.

The carrier sense multiple access (CSMA) technique used in IEEE 802.11 is intended to provide fair and equal access to all devices. It is essentially a "listen before talk" (LBT) mechanism. When networks become overloaded, the performance becomes uniformly poor for all users and all types of data. QoS modifies the access rules to provide a useful form of "controlled unfairness." Data that are identified as having a higher priority are given preferential access to the medium. They will, therefore, gain access at the expense of the lower priority traffic. IEEE P802.11e (i.e., a draft IEEE standard) defines changes to the operation of the IEEE 802.11 MAC to enable prioritization and classes of service over a WLAN.

---

[1]   This chapter was contributed by Tim Godfrey and is based on Draft 11.0 of IEEE P802.11e.

IEEE P802.11e: WHAT'S IN AND WHAT'S OUT

The Scope of IEEE 802.11 standard

The IEEE 802.11 standard is limited to specifying the MAC and PHY. QoS is a system-level concept and may involve higher layers to provide end-to-end QoS services. IEEE P802.11e specifies default settings that will provide effective prioritization of traffic in the MAC and PHY only, based on the priorities passed through the MAC service access point (SAP). Higher layer protocols that permit path establishment and bandwidth reservation may be used to allow applications needing QoS to set traffic specifications (TSPECs). These higher layer protocols may interface with the MAC and PHY to provide appropriate QoS parameters at those layers.

The use of higher layer protocols with IEEE P802.11e is supported, but not standardized. The IEEE 802.11 Working Group may decide to publish recommended practices to indicate uses of higher layer protocols for specific QoS applications and scenarios. Alternately, industry groups such as Wi-Fi Alliance could provide these recommendations.

Mandatory and optional features

Implementation of IEEE P802.11e is optional because IEEE P802.11e is an extension to the IEEE 802.11 base standard. The IEEE P802.11e QoS functions are designed to operate in a mixed environment with some QoS-capable STAs (QSTAs) and some non-QoS-capable (i.e., legacy) STAs.

Within IEEE P802.11e, there are several additional optional features, which may be implemented in an IEEE P802.11e device: block acknowledgment, automatic power save delivery (APSD), and direct link setup (DLS). These features are detailed later in this chapter.

Limits of WLANs

The task of managing QoS on a WLAN is difficult for many reasons. First, the limits of radio design, spectrum, and power cause the available bandwidth of a WLAN to be orders of magnitude slower than IEEE 802.3 wired Ethernet networks at any point in time. When WLANs are interconnected with wired

LANs, the WLAN will be the bandwidth bottleneck, and the QoS functions will be stressed from the outset. On top of that, WLANs operate in a very challenging environment. The medium is several orders of magnitude less reliable than copper or optical networks. Conditions change rapidly. WLAN connections may have to reduce their data rate, or connections may be lost altogether. While IEEE P802.11e provides mechanisms to tightly control the use of available wireless bandwidth, there is no such thing as guaranteed QoS over a WLAN connection.

Background of the legacy IEEE 802.11 MAC

The IEEE 802.11 base standard defines two channel access mechanisms, called *coordination functions*. These coordination functions determine when a STA is permitted to transmit and when it must be prepared to receive data. The fundamental mechanism is the distributed coordination function (DCF), which uses the CSMA mechanism. While wired networks implement collision detection on the medium (CSMA/CD), the nature of the wireless medium precludes a reliable collision detection mechanism.

WLANs use collision avoidance (CSMA/CA). The basic rule of CSMA is LBT. The PHY provides a clear channel assessment (CCA) signal to the MAC to indicate whether any other STAs are transmitting. IEEE 802.11 includes a virtual carrier sense mechanism to reduce problems with hidden STAs. This mechanism is supplemented in IEEE P802.11e, but will not be detailed further.

The second coordination function is point coordination function (PCF), which is discussed on the next page.

To keep multiple STAs from attempting to transmit simultaneously when another STA stops transmitting, a random backoff is used. A STA wanting to transmit will generate a random number within a range called the *contention window* (CW). Once the STA senses that the medium is unoccupied for a short time, it starts counting down from the random backoff. If the count reaches zero, it starts transmitting. If another STA has selected a lower number and starts transmitting first, the STA will hold its current backoff count and resume counting when the medium becomes clear again after the intervening transmission.

The IEEE 802.11 standard defines a number of frame exchange sequences, which can continue without medium contention. The CSMA rules cause a frame exchange sequence to look like continuous activity on the medium. At the end of frame exchange sequences, STAs must sense the medium and perform backoff before transmitting again.

The IEEE 802.11 standard also defines an optional PCF. See Figure 5–1. This function allows different access rules based on polling by a point coordinator (PC) operating at an AP. The PC uses special control frames to divide the time between Beacon frames into a contention-free period (CFP) and a contention period (CP). During the CFP, the AP maintains control of the medium by polling each STA that has requested to be on the polling list and accepting a frame in response to the poll. There is a continuing exchange of frames going to and from the AP, with no time spent for backoff or contention. Because STAs transmit only when polled, there are no collisions. After the AP has completed the polling of all the STAs, the CP begins, and access using the DCF rules resumes.

Implementation of the PCF is an option, and it has not been widely implemented. Although the PCF appears to have the potential to deliver QoS, there are a number of limitations. The hybrid coordination function (HCF) defined in IEEE P802.11e provides all the advantages and capabilities of the PCF with none of the disadvantages and limitations.

FUNDAMENTALS OF IEEE P802.11e OPERATION

Hybrid coordination function (HCF)

The HCF replaces the legacy DCF and PCF in a STA implementing IEEE P802.11e (i.e., the QSTA). The HCF is mandatory in all QSTAs. Within the HCF, there are two access mechanisms: enhanced distributed channel access (EDCA) and hybrid controlled channel access (HCCA). Unlike the legacy PCF, which used different frame exchange sequences in the CP and the CFP, the HCF defines a uniform set of frame exchange sequences that are usable at any time. In summary: HCF = EDCA + HCCA. See Figure 5–2.

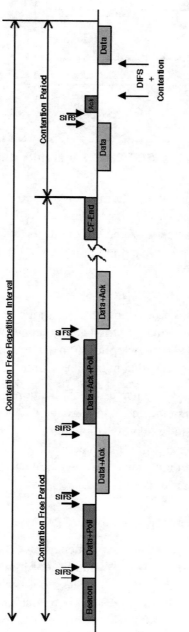

Figure 5–1: Summary of IEEE 802.11 coordination functions during the optional CFP and the CP

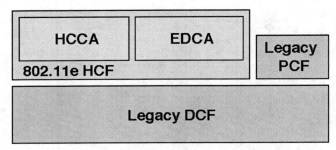

Figure 5–2: IEEE P802.11e architecture

The HCF allocates the right to transmit through transmit opportunities (TXOPs) granted to QSTAs. A STA may obtain a TXOP through one or both of the channel access mechanisms. A TXOP grants a particular QSTA the right to use the medium at a defined point in time, for a defined maximum duration. The allowed duration of TXOPs are communicated globally in the Beacon frame for STAs using EDCA.

The HCF introduces new acknowledgment rules. In the IEEE 802.11 base standard, every unicast data frame required an immediate response using an ACK frame. The HCF adds two new options: no acknowledgment and block acknowledgment. These are specified in the QoS Control field of data frames.

The no acknowledgment option is useful for applications with very low jitter tolerance, such as streaming multimedia, where the data would not be useful after the delay of a retry. Block acknowledgments increase efficiency by aggregating the ACK frames for multiple received frames into a single response. The implementation of block acknowledgment is optional and is discussed in more detail later in this chapter.

Enhanced distributed channel access (EDCA)

EDCA contention access is an extension of the legacy CSMA/CA DCF mechanism to include priorities. The CW and backoff times are adjusted to change the probability of gaining medium access to favor higher priority classes. A total of eight user priority (UP) levels are available. Each priority is mapped to an access category (AC), which corresponds to one of four transmit queues. See Figure 5–3.

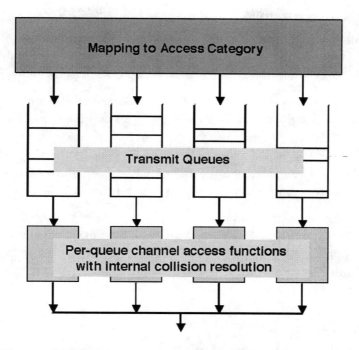

**Figure 5–3: EDCA queue architecture with four queues
(i.e., priority levels), to which eight UP levels are mapped
according to a fixed mapping**

Each queue provides frames to an independent channel access function, each of which implements the EDCA contention algorithm. When frames are available in multiple transmit queues, contention for the medium occurs both internally and externally, based on the same coordination function. Therefore, the internal scheduling resembles the external scheduling. Internal collisions are resolved by allowing the frame with higher priority to transmit, while the lower priority invokes a queue-specific backoff as if a collision had occurred.

The parameters defining EDCA operation, such as the minimum idle delay before contention and the minimum and maximum CWs, are stored locally at the QSTA. These parameters will be different for each AC and can be dynamically updated by the QoS AP (QAP) for each AC through the EDCA parameter sets. These EDCA parameter sets are sent from the QAP as part of the Beacon frame and in Probe and Reassociation Response frames. This adjustment allows the STAs in the network to adjust to changing conditions and gives the QAP the ability to manage overall QoS performance.

Under EDCA, STAs and APs use the same access mechanism and contend on an equal basis at a given priority. A STA that wins an EDCA contention is granted a TXOP – the right to use the medium for a period of time. The duration of this TXOP is specified per AC and is contained in the TXOP Limit field of the AC parameter record in the EDCA parameter set. A QSTA can use a TXOP to transmit multiple frames within an AC.

If the frame exchange sequence has been completed and there is time remaining in the TXOP, the QSTA may extend the frame exchange sequence by transmitting another frame in the same AC. The QSTA must ensure that the transmitted frame and any necessary acknowledgment can fit into the time remaining in the TXOP.

EDCA admission control

Contention-based medium access is susceptible to severe performance degradation when overloaded. In overload conditions, the CWs become large, and more and more time is spent in backoff delays rather than sending data. Admission control regulates the amount of data contending for the medium.

EDCA admission control is mandatory at the AP, and optional at the STA. The AP may indicate that it requires STAs to support admission control and explicitly request access rights if they wish to use an AC.

Admission control is negotiated by the use of a TSPEC. A STA specifies its traffic flow requirements (e.g., data rate, delay bounds, packet size) and requests the QAP to create a TSPEC by sending the Add TSPEC (ADDTS) management frame. The QAP calculates the existing load based on the current set of issued TSPECs. Based on the current conditions, the QAP may accept or deny the new TSPEC request. If the TSPEC is denied, the high-priority AC inside the QSTA is not permitted to use the high-priority access parameters, but it must use lower priority parameters instead. Admission control is not intended to be used for the "best effort" and "background" traffic classes.

HCF controlled channel access (HCCA)

The HCCA mechanism uses a hybrid coordinator (HC) to centrally manage medium access. The intent of HCCA is to increase efficiency by reducing the contention on the medium.

The HC has privileged access to the medium because it can initiate a transmission after waiting a shorter time than the shortest backoff delay of any STA using EDCA. Under control of the HC, a nearly continuous sequence of frame exchanges can be maintained with short, fixed delays between frames. The interframe delay does not increase with increasing traffic, unlike the CW used in EDCA access. There is no possibility of collisions, except from STAs on the same frequency that are not under control of the HC. HCCA supports parameterized QoS, where specific QoS flows from applications can have individually tailored QoS parameters and have tighter control of latency and scheduling.

IEEE P802.11e defines new QoS frame types that allow the HC to send any combination of data, poll, and acknowledgment to a STA in a single frame. When the HC sends a poll to a QSTA, the QoS Control field contains a TXOP limit value that specifies the duration of the granted TXOP.

The HC is responsible for controlling the allocation of time on the medium through the use of polled TXOPs. The HC is guided in its decisions on TXOP allocation through the use of TTSPECs and the associated STAs queue size information. As previously described, TSPECs define the QoS parameters required by a QSTA for a specific traffic class. TSPECs are requested by the QSTA, and the QAP may grant or deny a TSPEC request. The handling of TSPECs is described later in this chapter.

IEEE P802.11e FRAME FORMATS

As a backward-compatible option in the IEEE 802.11 standard, IEEE P802.11e extends the existing frame formats and adds new frame subtypes and fields that were previously reserved.

Within the Frame Control field of the MAC header, the definitions of frame types and subtypes are the same as in the IEEE 802.11 standard (see Figure 3–6 and Table 3–1 in this handbook).

However, IEEE P802.11e adds to the control, data, and management frame subtypes to support the new QoS functionality.

New control frame subtypes

IEEE P802.11e expands Table 1 from the IEEE 802.11 standard to include two new subtypes for control frames (see Table 5–1).

Table 5–1: New valid type and subtype combinations for control frames

Type value b3 b2	Type description	Subtype value b7 b6 b5 b4	Subtype description
01	Control	1000	BlockAckReq
01	Control	1001	BlockAck

The frame format of the BlockAckReq frame is defined in Figure 5–4.

- The Duration field is consistent with the normal IEEE 802.11 usage and is set to the remaining time in the frame exchange sequence (i.e., the ACK frame or BlockAck frame sent in response, plus a SIFS interval)

- The BAR Control field identifies the traffic identifier (TID) of the requested block acknowledgment.

- The Starting Sequence Control field identifies the specific MSDU for which the block acknowledgment is being requested. This value is the number of the first of a block of up to 64 MSDUs that may be acknowledged in the block acknowledgment response.

Figure 5–4: BlockAckReq frame format

The frame format of the BlockAck frame is defined in Figure 5–5.

- The BA Control field indicates whether immediate or delayed mode is being used for the block acknowledgment. If immediate block acknowledgment mode is being used, the BlockAck frame is sent in response to a BlockAckRequest frame (after a SIFS). If delayed block acknowledgment mode is being used, the BlockAck frame is sent independently, and the BlockAck frame is acknowledged with a regular ACK frame.

- The Block ACK Starting Sequence Control field contains a sequence number, which is set to the same value as in the immediately previously received BlockAckReq frame.

- The Block ACK Bitmap field indicates the acknowledgment of up to 64 MSDUs, starting with the MSDU indicated in the sequence number. Each of the 64 MSDUs is represented by a 16-bit field to support the 16 possible fragments (if fragmentation is used). As a result, 64 MSDUs * 16 bits gives 1024 bits, or 128 octets.

New data frame subtypes

IEEE P802.11e expands Table 1 from the IEEE 802.11 standard to include eight new subtypes of data frames (see Table 5–2).

IEEE P802.11e defines the data subtypes with b7 set to one to indicate QoS data. These subtypes were previously all reserved. The functions of the QoS data subtypes 8 through 15 correspond to the non-QoS data subtypes 0 through 7 in the IEEE 802.11 standard.

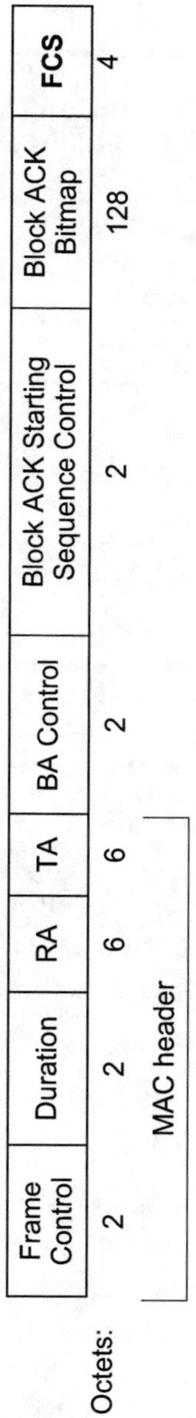

Frame Control	Duration	RA	TA	BA Control	Block ACK Starting Sequence Control	Block ACK Bitmap	**FCS**
2	2	6	6	2	2	128	4

Octets:

MAC header

Figure 5–5: BlockAck frame format

Table 5–2:   New valid type and subtype combinations of data frames

Type value b3 b2	Type description	Subtype value b7 b6 b5 b4	Subtype description
10	Data	1000	QoS Data
10	Data	1001	QoS Data + CF-ACK
10	Data	1010	QoS Data + CF-Poll
10	Data	1011	QoS Data + CF-ACK + CF-Poll
10	Data	1100	QoS Null (no data)
10	Data	1101	Reserved
10	Data	1110	QoS CF-Poll (no data)
10	Data	1111	QoS CF-Poll + CF-ACK (no data)

QoS Control field

IEEE P802.11e adds a 2-octet field, QoS Control, to the MAC header of QoS data frames. The QoS Control field occurs immediately after the Address 4 field (if used) and immediately preceding the frame body as shown in Figure 5–6.

Table 5–3 defines the subfield definitions for the QoS Control field. The subfields have minor differences depending on the type of frame.

The TID subfield indicates the traffic category or traffic stream of the data in the frame, except in the case of a QoS + CF-Poll frame (no data). This subfield has 16 possible TID values, which are divided into two groups: values of 0–7 identify traffic categories and 8–15 identify traffic streams. In the case of QoS + CF-Poll frames (no data), the TID is a nonbinding suggestion to the STA of the TID for which the poll is intended.

Frame Control	Duration/ ID	Address 1	Address 2	Address 3	Sequence Control	Address 4	QoS Control	Frame Body	FCS
2	2	6	6	6	2	6	2	N	4

Octets:

MAC header

Figure 5–6: MAC frame format

Table 5–3:   QoS Control field format

Applicable data frame	Bits 0-3	Bit 4	Bits 5-6	Bit 7	Bits 8-15
QoS (+)CF-Poll frames sent by HC	TID	EOSP	ACK Policy	Reserved	TXOP limit in units of 32 µs
QoS Data, QoS Null (no data), QoS Data + CF-Ack, and QoS CF-Ack frames sent by HC	TID	EOSP	ACK Policy	Reserved	QAP PS Buffer State
QoS data frames sent by non-AP QSTAs	TID	0	ACK Policy	Reserved	TXOP duration requested in units of 32 µs
	TID	1	ACK Policy	Reserved	Queue size in units of 256 octets

The EOSP bit allows the HC to indicate to a STA that the current frame is the end of the service period. Service periods are established through the TSPEC negotiation process, and they allow a STA to go into a low-power state except during the scheduled service period. If the HC knows that the service period will end prior to the scheduled time, the EOSP bit allows the STA to enter the low-power state immediately.

The ACK Policy subfield is defined as shown in Table 5–4.

The TXOP subfield has several roles depending, first, on whether the frame is sent by the HC or by a non-AP QSTA. The first two rows of Table 5–3 [QoS Control field format] are for frames sent by the HC to a QSTA. If the frame is a poll, the HC conveys the TXOP limit it is granting to the polled STA. For other frame types sent by the HC, the TXOP limit is not relevant and thus reserved. Frames sent from a non-AP QSTA may convey either a requested TXOP duration or the current queue size of the TID, depending on the setting of bit 4.

Table 5–4:   ACK Policy subfield of the QoS Control field

Bit 5	Bit 6	Meaning
0	0	Normal acknowledgment. The addressed recipient returns an ACK or QoS + CF-Ack frame after a SIFS period, according to the procedures defined in 9.2.8, 9.3.3, and 9.9.2.3 in IEEE P802.11e. The ACK Policy subfield is set to this value in all directed frames in which the sender requires acknowledgment. For QoS Null (no data) frames, this value is the only permissible value for the ACK Policy subfield.
1	0	No acknowledgment. The addressed recipient takes no action upon receipt of the frame. More details are provided in 9.11 in IEEE P802.11e. The ACK Policy subfield is set to this value in all directed frames in which the sender does not require acknowledgment. This combination is also used for broadcast and multicast frames that use the QoS data frame format.
0	1	No explicit acknowledgment. There may be a response frame to the frame that is received, but it is neither the ACK frame nor any data frame of subtype + CF-Ack. For QoS CF-Poll (no data) frames and QoS CF-Ack + CF-Poll (no data) frames, this value is the only permissible value for the ACK Policy subfield.
1	1	Block acknowledgment. The addressed recipient takes no action upon the receipt of the frame except for recording the state. The recipient can expect a BlockAckReq frame in the future, to which it responds using the procedure described in 9.10 in IEEE P802.11e.

The QAP PS Buffer State subfield allows the HC to indicate its power save buffer state to a STA. See Figure 5–7.

B8	B9	B10	B11	B12 B13 B14 B15
Reserved	Buffer State Indicated	Highest Priority Buffered AC		QAP Buffered Load

Figure 5–7: QAP PS Buffer State subfield format

- The Buffered State Indicated subfield is 1 bit in length and is used to indicate whether the QAP PS buffer state is specified. A value of one indicates that the QAP PS buffer state is specified.

- The Highest Priority Buffered AC subfield is 2 bits in length and is used to indicate the AC of the highest priority traffic remaining that is buffered at the QAP, excluding the MSDU of the present frame.

- The QAP Buffered Load subfield is 4 bits in length and is used to indicate the total buffer size, rounded up to the nearest multiple of 4096 octets and expressed in units of 4096 octets, of all MSDUs buffered at the QAP (excluding the MSDU of the present QoS data frame). A QAP Buffered Load subfield value of 15 indicates that the buffer size is greater than 57 344 octets. A QAP Buffered Load subfield value of zero is used solely to indicate the absence of any buffered traffic for the indicated highest priority buffered AC when the Buffer State Indicated bit is one.

- When the Buffered State Indicated subfield is set to zero, the Highest Priority Buffered AC subfield and the QAP Buffered Load subfield are reserved; and the values of these subfields are either unspecified or unknown.

New extensions to management frames

Existing management frames, such as Beacon, Association Request, Association Response, Probe Request, and Probe Response, are extended by IEEE P802.11e with new information elements.

Also, the Action [management] frame is more fully developed in IEEE P802.11e. The Action frame is a general format, defined and used in IEEE P802.11e and IEEE 802.11h in a consistent manner. The specific Action frame formats defined in IEEE P802.11e are described below.

The information elements defined in the IEEE 802.11 standard are extended in IEEE P802.11e to add new QoS functionality to existing management frame formats.

Beacon [management] frames

Table 5 in the IEEE 802.11 standard defines a set of fixed fields and information elements for Beacon frames.

IEEE P802.11e adds the information elements listed in Table 5–5 for Beacon frames.

Table 5–5:   QoS-related information elements for Beacon frames

Order	Information	Notes
14	QBSS Load	The QBSS Load information element is present only within Beacon frames generated by QAPs. The QBSS Load information element is present when dot11QoSOptionImplemented and dot11QBSSLoadImplemented are both true.
15	EDCA Parameter Set	The EDCA Parameter Set information element is present only within Beacon frames generated by QAPs. The EDCA Parameter Set information element is present when dot11QoSOptionImplemented is true and the QoS Capability information element is not present.
23	QoS Capability	The QoS Capability information element is present only within Beacon frames generated by QAPs. The QoS Capability information element is present when dot11QoSOptionImplemented is true and EDCA Parameter Set information element is not present.

Capability Information field

The Capability Information field in the frame body of Beacon frames is a fixed field, which is extended as shown in Figure 5–8. The bits defined in IEEE P802.11e are as follows:

- The QoS bit indicates that the (optional) QoS facility is supported.

- The APSD bit indicates that automatic power save delivery is supported.

- The Delayed Block ACK bit indicates that delayed block acknowledgments are supported.

- The Immediate Block ACK bit indicates that immediate block acknowledgments are supported.

If the QoS Capability bit is set by a STA, the CF-Pollable and CF-Poll Request bits must be set to zero. When used in conjunction with the QoS Capability bit, the CF-Pollable and CF-Poll Request bits are defined as shown in Table 5–6 for a HC at an AP.

Table 5–6:   AP usage of QoS Capability, CF-Pollable and CF-Poll Request

QoS	CF-Pollable	CF-Poll Request	Meaning
1	0	0	QAP (HC) does not use CFP for delivery of unicast data frames
1	0	1	QAP (HC) uses CFP for delivery, but does not send CF-polls to non-QoS STAs
1	1	0	QAP (HC) uses CFP for delivery and sends CF-polls to non-QoS STAs
1	1	1	Reserved

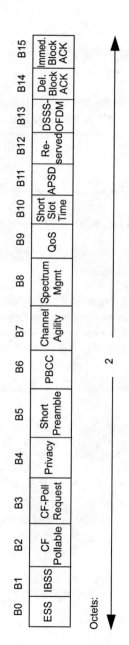

Figure 5–8: Capability Information field format

Association Request frames

IEEE P802.11e adds the information elements listed in Table 5–7 to Association Request frames.

Table 5–7: QoS-related information elements for Association Request frames

Order	Information
9	QoS Capability

(Re)Association Response frames

IEEE P802.11e adds the information elements listed in Table 5–8 to (Re)Association Response frames.

Table 5–8: QoS-related information elements for (Re)Association Response frames

Order	Information
5	EDCA Parameter Set

Reassociation Request frames

IEEE P802.11e adds the information elements listed in Table 5–9 to Reassociation Request frames.

Table 5–9: QoS-related information elements for Reassociation Request frames

Order	Information
10	QoS Capability

Probe Response frames

IEEE P802.11e adds the information elements listed in Table 5–10 to Probe Response frames.

Table 5–10: QoS-related information elements for Prove Response frames

Order	Information	Notes
13	QBSS Load	The QBSS Load information element is present when dot11QoSOptionImplemented and dot11QBSSLoadImplemented are both true. The QBSS Load information element is always present within Probe Response frames generated by QAPs.
14	EDCA Parameter Set	The EDCA Parameter Set information element is present when dot11QoSOptionImplemented is true. The EDCA Parameter Set information element is always present within Probe Response frames generated by QAPs.

Action [management] frames

IEEE 802.11h introduced the management frame of subtype Action. Table 1 in the IEEE 802.11 standard was expanded to include this subtype of management frame (see Table 5–11).

Table 5–11: Valid type and subtype combinations of management frames

Type value b3 b2	Type description	Subtype value b7 b6 b5 b4	Subtype description
00	Management	1101	Action

The frame body of an Action frame contains an Action field, which consists of a Category subfield and a variable-length Action Details subfield, as shown in Figure 5–9.

For QoS, DLS, and block acknowledgment functionality, the Category subfield contains the corresponding value shown in Table 5–12.

Table 5–12: Category subfield values

Name	Value	Note
Spectrum management	0	Defined in IEEE 802.11h
QoS	1	Defined in IEEE P802.11e
DLS	2	Defined in IEEE P802.11e
Block ACK	3	Defined in IEEE P802.11e

Several Action frame formats are defined for QoS functionality. The Action Details subfield differentiates the formats based on the categories defined in IEEE P802.11e. This subfield always contains an Action subfield and also contains other subfields that vary depending on the frame function. For example, action request and response frames may contain a Dialog Token subfield.

A dialog token is used to match action responses with action requests when multiple requests are outstanding. The dialog token is an arbitrary number chosen by the sender. The action response uses the same dialog token value as the action request to which it is responding.

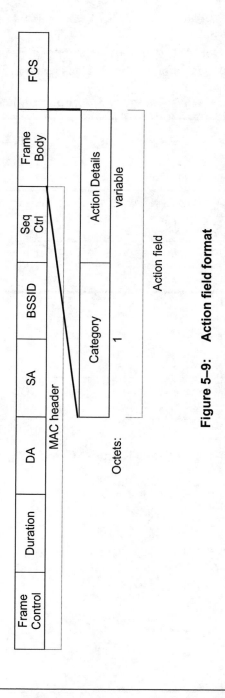

Figure 5–9: Action field format

QoS Action frames

The frame bodies of QoS Action frames are shown in Figure 5–10 through Figure 5–13 based on the Action subfield values given in Table 5–13.

Table 5–13:   QoS (Category 1) Action subfield values

Action subfield value	Description	Frame format
0	ADDTS request	See Figure 5–10
1	ADDTS response	See Figure 5–11
2	DELTS	See Figure 5–12
3	Schedule	See Figure 5–13
4–255	Reserved	—

	Category (1)	Action (0)	Dialog Token	TSPEC	TCLAS (optional)	TCLAS Processing (optional)
Octets	1	1	1	48	variable	3

Figure 5–10:   ADDTS Request frame body format

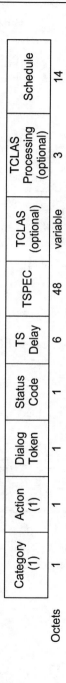

Figure 5–11:   ADDTS Response frame body format

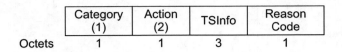

Figure 5–12: DELTS frame body format

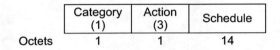

Figure 5–13: Schedule frame body format

DLS Action frames

The frame bodies of DLS Action frames are shown in Figure 5–14 through Figure 5–16 based on the Action subfield values given in Table 5–14.

Table 5–14: DLP (Category 2) Action subfield values

Action subfield value	Description	Frame format
0	DLS Request	See Figure 5–14
1	DLS reSponse	See Figure 5–15
2	DLS Teardown	See Figure 5–16
3–255	Reserved	—

In a DLS request frame, the DLS Timeout Value subfield value is 16 bits long and indicates a duration in seconds. If there are no direct link frame exchanges in this period, the direct link is terminated.

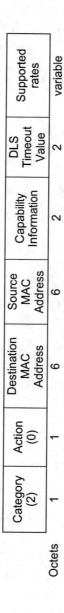

Figure 5–14: DLS Request frame body format

Category (2)	Action (1)	Status Code	Destination MAC Address	Source MAC Address	Capability Information	Supported rates
1	1	1	6	6	2	variable

Octets

Figure 5–15: DLS Response frame body format

	Category (2)	Action (2)	Destination MAC Address	Source MAC Address	Reason Code
Octets	1	1	6	6	1

Figure 5–16:   DLS Teardown frame body format

Block ACK Action frames

The frame bodies of Block ACK Action frames are shown in Figure 5–17, Figure 5–19, and Figure 5–20 based on the Action subfield values given in Table 5–15.

Table 5–15:   Block ACK (Category 3) Action subfield values

Action subfield values	Description	Frame format
0	ADDBA Request	See Figure 5–17
1	ADDBA Response	See Figure 5–19
2	DELBA	See Figure 5–20
3-255	Reserved	—

	Category (3)	Action (0)	Dialog Token	Block ACK Parameter Set	Block ACK Timeout Value	Block ACK Starting Sequence Control
Octets	1	1	1	2	2	2

Figure 5–17:   ADDBA Request frame body format

The Block ACK Parameter Set subfield of an ADDBA request frame is a fixed field, and its value indicates the type of ACK policy requested (i.e., immediate or delayed), the TID for which the block acknowledgment is requested, and the number of buffers available for holding received frames before the block acknowledgment is generated. See Figure 5–18.

B0	B1	B2 B5	B6 B15
Reserved	Block ACK Policy	TID	Buffer Size

Octets: 2

Figure 5–18:   Block ACK Parameter Set subfield format

The Block ACK Timeout Value subfield of an ADDBA request frame is 2 bytes long, and its value indicates the number of seconds of inactivity allowed before the block acknowledgment is considered inactive.

The Block ACK Starting Sequence Control subfield of an ADDBA request frame is the same as defined in a corresponding BlockAckReq [control] frame. It defines the sequence number of the first MSDU for which that BlockAckReq frame is sent.

Category (3)	Action (1)	Dialog Token	Status code	Block ACK Parameter Set	Block ACK Timeout Value
1	1	1	1	2	2

Octets

Figure 5–19:   ADDBA Response frame body format

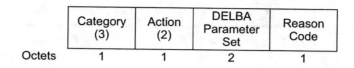

Figure 5–20: DELBA frame body format

The DELBA frame is used to terminate an existing block acknowledgment. The DELBA Parameter Set subfield is a fixed field and contains the Initiator subfield and TID subfield (see Figure 5–21). The Initiator bit is set to one if the initiator of the block acknowledgment is requesting the termination. The TID is the same as the one used to initiate the block acknowledgment.

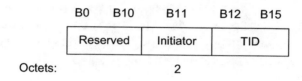

Figure 5–21: DELBA Parameter Set subfield

New information element formats

IEEE P802.11e defines the information elements shown in Table 5–16. Five of these new information elements warrant further elaboration.

Table 5–16: Information elements defined by IEEE P802.11e and their associated element IDs

Information element	Element ID
QBSS Load	11
EDCA Parameter Set	12

Table 5–16: Information elements defined by IEEE P802.11e and their associated element IDs *(Continued)*

Information element	Element ID
TSPEC	13
Traffic Classification	14
Schedule	15
TS Delay	43
TCLAS Processing	44
QoS Capability	46

QBSS Load information element

The QBSS Load information element contains information on the current STA population and traffic levels in the QBSS. It is used in Beacon frames and Probe Response frames. It provides information to STAs to help them make decisions on where to associate and when to roam to another AP. See Figure 5–22.

Element ID (11)	Length (5)	Station Count	Channel Utilization	Available Admission Capacity

Octets:	1	1	2	1	2

Figure 5–22: QBSS Load information element format

The Channel Utilization field value indicates the percentage of time the QAP sensed the medium as busy. The Available Admission Capacity field value is the QAP's indication of the amount of time available to be allocated to new streams via admission control.

EDCA Parameter Set information element

The EDCA Parameter Set information element provides information needed by QSTAs for proper operation during the CP. It allows the QAP to dynamically adjust EDCA parameters used by STAs for contention within the BSS, as appropriate for traffic and other conditions. See Figure 5–23.

The QoS Info field of the EDCA Parameter Set information element is defined as shown in Figure 5–24 and Figure 5-25.

The subfields of the QoS Info field are defined as follows:

- The EDCA Parameter Set Update Count subfield increments its counter whenever the EDCA parameter set values change. This information allows a STA that has been asleep to determine whether it has missed any updates. The updated EDCA parameter set can be requested by using a Probe Request frame if necessary.

- The Q-Ack subfield value indicates whether the sender supports non-piggybacked acknowledgments. If Q-Ack is supported, a STA must be able to process an acknowledgment in the form of a QoS (+)Data + CF-Ack frame addressed to a different STA. The QoS Info field is present in the Beacon frame sent by QAPs and in the Probe Response frame sent by QSTAs.

- The Queue Request bit is used by QAPs to indicate whether they can process the Queue Size subfield in the QoS Control field.

- The TXOP Request bit is used by QAPs to indicate whether they can process the TXOP Request subfield in the QoS Control field.

- The UPSD flags are set to indicate that the corresponding access category (AC) is trigger-enabled and delivery-enabled.

- The MAX SP Length subfield indicates the maximum number of downlink unicast frames the QAP may deliver to a non-AP QSTA during any service period triggered by the non-AP QSTA. See Table 5–17.

- The More Data ACK bit is used by non-AP QSTAs to indicate whether they can process ACK frames with the More Data bit in the Frame Control field set to one and will interpret the bit being set to one as an instruction to remain in the awake state.

Element ID (12)	Length (18)	QoS Info	Reserved	AC_BE Parameters Record	AC_BK Parameters Record	AC_VI Parameters Record	AC_VO Parameters Record
1	1	1	1	4	4	4	4

Octets:

Figure 5–23:   EDCA Parameter Set information element format

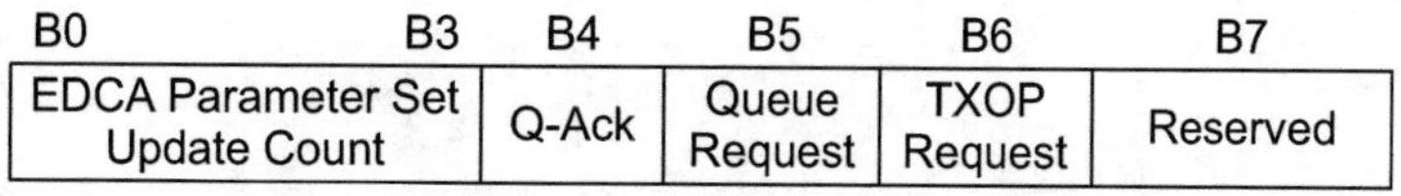

B0	B3	B4	B5	B6	B7
EDCA Parameter Set Update Count		Q-Ack	Queue Request	TXOP Request	Reserved

Figure 5–24: QoS Info field when sent by a QAP

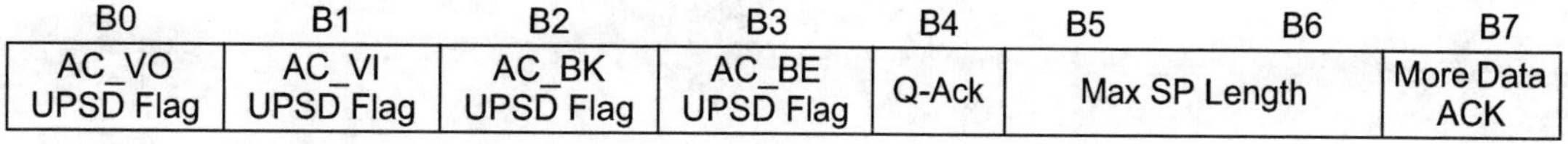

B0	B1	B2	B3	B4	B5	B6	B7
AC_VO UPSD Flag	AC_VI UPSD Flag	AC_BK UPSD Flag	AC_BE UPSD Flag	Q-Ack	Max SP Length		More Data ACK

Figure 5–25: QoS Info field when sent by a non-AP QSTA

Table 5–17: MAX SP Length subfield values

Bit 5	Bit 6	Usage
0	0	QAP may release all buffered frames
1	0	QAP may release a maximum of two frames per service period
0	1	QAP may release a maximum of four frames per service period
1	1	QAP may release a maximum of six frames per service period

The format of the four AC parameter record fields of the EDCA Parameter Set information element is shown in Figure 5–26.

ACI/AIFSN	ECWmin/ ECWmax	TXOP Limit
Octets: 1	1	2

Figure 5–26:   AC_BE/AC_BK/AC_VI/AC_VO Parameters Record field format

The ACI/AIFSN subfield of the AC parameters record fields is shown in Figure 5–27, and its subfields are defined after the figure.

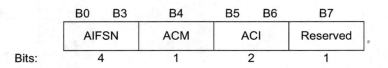

B0 B3	B4	B5 B6	B7
AIFSN	ACM	ACI	Reserved
Bits: 4	1	2	1

Figure 5–27:   ACI/AIFSN subfield format

- The Arbitration Interframe Space Number (AIFSN) subfield indicates the number of slots after a SIFS duration that a non-AP QSTA should defer before either invoking a backoff or starting a transmission.

- The Access Category Index (ACI) subfield is mapped to ACs as shown in Table 5–18.

Table 5–18: ACI-to-AC coding

ACI	AC abbreviation	AC name
00	AC_BE	Best Effort
01	AC_BK	Background
10	AC_VI	Video
11	AC_VO	Voice

- The Admission Control Mandatory (ACM) bit indicates whether admission control is mandatory for a STA to use the AC.

The ECWmin and ECWmax subfields of the AC parameters record fields indicate the minimum and maximum CW values, respectively, that are to be used for EDCA access for the AC.

The TXOP Limit subfield of the AC parameters record fields indicates the length of the TXOP that is granted by winning an EDCA contention. A value of zero indicates one MSDU, the same behavior as allowed in the IEEE 802.11 base standard.

TSPEC information element

The TSPEC information element defines the characteristics and expectations of a traffic flow or stream. The list of parameters is more extensive than is typically needed for any particular flow, but parameters that are unused may be set to zero. See Figure 5–28.

The TS Info field of the TSPEC information element is defined as shown in Figure 5–29, and its subfields are defined after the figure.

Element ID (13)	Length (55)	TS Info	Nominal MSDU Size	Maximum MSDU Size	Minimum Service Interval	Maximum Service Interval	Inactivity Interval	Suspension Interval
1	1	3	2	2	4	4	4	4

Service Start Time	Minimum Data Rate	Mean Data Rate	Peak Data Rate	Maximum Burst Size	Delay Bound	Minimum PHY Rate	Surplus Bandwidth Allowance	Medium Time
4	4	4	4	4	4	4	2	2

Octets:

Figure 5–28: TSPEC information element format

Bits:	Traffic Type	TSID	Direction	Access Policy	Aggregation	APSD	User Priority	TSInfo ACK Policy	Schedule	Reserved
	1	4	2	2	1	1	3	2	1	7
	B0	B1 B4	B5 B6	B7 B8	B9	B10	B11 B13	B14 B15	B16	B17 B23

Figure 5–29: TS Info field

- The Traffic Type subfield is set to one for isochronous traffic at a fixed repetition rate or set to zero for aperiodic traffic.

- The TSID subfield creates the association between the TID subfield in the QoS Control field and the corresponding TSPEC. TID values between 8 and 15 are treated as traffic stream identifiers (TSIDs) and are mapped to TSPECs.

- The Direction subfield indicates whether the TSPEC is an uplink (QSTA to QAP), downlink (QAP to QSTA), or direct link (QSTA to QSTA). Bi-directional links mean that two flows, one in each direction, are created with the same parameters.

- The Access Policy subfield selects whether the TSPEC is to use EDCA (bit 7), HCCA (bit 8), or both.

- The Aggregation subfield indicates whether the QAP must combine delivery and polling of all schedules to a particular QSTA into a single service period. Aggregation allows the QSTA to sleep for the remainder of the time.

- The APSD subfield indicates that automatic power save delivery is to be used for the traffic associated with the TSPEC.

- The User Priority subfield indicates the actual value of the user priority to be used for the transport of MSDUs belonging to this traffic stream in cases where relative prioritization is required.

- The TSInfo ACK Policy subfield indicates whether the traffic associated with this TID will use normal acknowledgment, no acknowledgment, or block acknowledgment.

- The Schedule bit determines the type of APSD used when the access policy is EDCA, and APSD is set. A one indicates scheduled APSD, and a zero indicates unscheduled APSD. Unscheduled APSD allows a QSTA to send a "trigger" frame to the QAP to initiate the unscheduled service period.

The Nominal MSDU Size field of the TSPEC information element specifies the size of MSDUs in the stream belonging to this TSPEC. If the frames will have a constant size, the fixed bit is set to one. If a variable size is expected, the fixed bit should be set to zero.

The Maximum MSDU Size field of the TSPEC information element specifies the maximum size, in octets, of MSDUs belonging to the traffic stream under this traffic specification

The Minimum Service Interval and Maximum Service Interval fields of the TSPEC information element specify the minimum interval, in microseconds, between the start of two successive service periods.

The Inactivity Interval field of the TSPEC information element specifies the maximum amount of time that may elapse without arrival or transfer of an MSDU belonging to the traffic stream before this traffic stream is deleted.

The Suspension Interval field of the TSPEC information element specifies the maximum amount of time that may elapse without arrival or transfer of an MSDU belonging to the traffic stream before polling is stopped for this traffic stream. A value of -1 disables the suspension interval.

The Service Start Time field of the TSPEC information element indicates the time when the service period starts. The service start time indicates to QAP the time when a non-AP QSTA first expects to be ready to send frames and a power-saving non-AP QSTA will be awake to receive frames.

The Minimum Data Rate, Mean Data Rate, and Peak Data Rate fields of the TSPEC information element specify the lowest, average, and peak data rates (at the MAC_SAP) for transport of MSDUs belonging to this traffic stream within the bounds of this traffic specification. These rates do not include the MAC and PHY overheads.

The Maximum Burst Size field of the TSPEC information element specifies the maximum burst, in octets, of the MSDUs belonging to this traffic stream that arrive at the MAC SAP at the peak data rate. A value of zero indicates that there are no bursts.

The Delay Bound field of the TSPEC information element specifies the maximum amount of time allowed to transport an MSDU belonging to the traffic stream in this TSPEC. The time is measured from originating MAC SAP to destination MAC SAP. The time includes the acknowledgment, if used.

The Minimum PHY Rate field of the TSPEC information element specifies the desired minimum PHY rate to use for this traffic stream, in bits per second, that is required for transport of the MSDUs belonging to the traffic stream in this TSPEC.

The Surplus Bandwidth Allowance field of the TSPEC information element specifies the excess allocation of time (and bandwidth) over and above the stated application rates required to transport an MSDU belonging to the traffic stream in this TSPEC. This field takes into account retransmissions, as the rate information does not include retransmissions. A value of one indicates that no additional allocation of time is requested.

The Medium Time field of the TSPEC information element contains the amount of time admitted to access the medium. This field is reserved in the ADDTS Request frame and is set by the HC in the ADDTS Response frame. The derivation of this field is described in H.2.2 if the IEEE 802.11 standard. This field is not used for controlled channel access.

Schedule information element

The Schedule information element is transmitted by the HC in the QAP to a non-AP QSTA. It announces the schedule that the HC/QAP follows for admitted streams originating from or destined to that non-AP QSTA in the future. The format of the Schedule element information is shown in Figure 5–30.

Element ID (15)	Length (14)	Schedule Info	Service Start Time	Service Interval	Maximum Service Duration	Specifi- cation Interval
1	1	2	4	4	2	2

Octets: (above values)

Figure 5–30:   Schedule information element format

The Schedule Info field of the Schedule information element is shown in Figure 5–31. These fields are defined the same as in the TS Info field of the TSPEC information element.

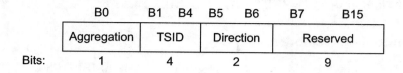

Figure 5–31:   Schedule Info field format

The Service Start Time field of the Schedule information element is defined the same as the Service Start Time field in the TSPEC information element.

The Service Interval field of the Schedule information element represents the measured time from the start of one service period to the start of the next service period.

The Maximum Service Duration field of the Schedule information element represents the longest time a service period will last.

The Specification Interval field of the Schedule information element specifies the time interval, in TUs (i.e., 1024 µs), to verify schedule conformance.

QoS Capability information element

The QoS Capability information element contains only the QoS Info field, which is defined the same as the QoS Info field of the EDCA Parameter Set information element. See Figure 5–32. The QoS Capability information element is used in Beacon frames and Probe Response frames when the full EDCA Parameter Set information element is not needed. Either the QoS Capability information element or the EDCA Parameter Set information element is used, but never are they used together.

Element ID (46)	Length (1)	QoS Info

Figure 5–32:   QoS Capability element format

OPTIONAL FEATURES IN IEEE P802.11e

Contention-free bursts (CFBs)

CFBs are not explicitly listed as an optional feature of IEEE P802.11e, but a QSTA or QAP may choose to use CFBs to improve efficiency by eliminating some contention. A CFB may be used when a QSTA or QAP has time remaining in a granted TXOP and additional data to send. Rather than contending for the medium again as would be required in legacy IEEE 802.11 devices, IEEE P802.11e allows a STA to resume transmitting after the short interframe space (SIFS) delay. A CFB is a special form of frame exchange sequence that fits within a TXOP. CFBs may be used during TXOPs that were gained under both EDCA and HCCA channel access functions.

IEEE P802.11e specifies methods for the QAP or QSTA to recover in the case of a transmission failure during a granted TXOP. Recovering means that the owner of the TXOP can reestablish operation before any other STA is able to access the medium through the normal contention mechanisms.

CFBs are also useful to improve throughput of IEEE 802.11g STAs in the presence of IEEE 802.11b devices, even if the network is not otherwise using IEEE P802.11e QoS mechanisms. The IEEE 802.11g STA is programmed to use a TXOP length comparable to the single frame duration of a STA using the legacy IEEE 802.11b rates. The IEEE 802.11g STA can send multiple frames during the CFB and occupy the medium for the same amount of time as a single IEEE 802.11b frame. The use of CFBs allows IEEE 802.11g devices to achieve the expected higher throughput by keeping IEEE 802.11b devices from taking a disproportionate amount of time on the medium.

Block acknowledgments

The legacy IEEE 802.11 MAC always sends an ACK frame after each frame that is successfully received. Block acknowledgment allows several data frames to be transmitted before an ACK frame is returned and increases the efficiency because every frame has a significant overhead for radio synchronization. Block acknowledgement is initiated through a setup and negotiation process between the QSTA and QAP. Once block acknowledgment has been established, multiple QoS data frames can be transmitted in aCFB with SIFS separation between the frames.

There are two block acknowledgment modes: immediate and delayed. When using the immediate mode, the sending STA transmits multiple data frames in aCFB separated by SIFS. The sender must obey the constraints of the TXOP duration within which it is currently operating. At the end of the burst, the sender transmits a BlockAckReq frame. The receiver must immediately respond with a BlockAck frame containing the acknowledgment status for the previous burst of data frames. See Figure 5–33.

The delayed mode allows the block acknowledgment to be sent in a subsequent TXOP following the burst. The same sequence of a CFB and BlockAckReq frame is used as in the immediate mode. In response to the BlockAckReq frame, the receiver simply sends a standard ACK frame, indicating that the block acknowledgment will be delayed. Delayed block acknowledgment increases the latency and is provided to support lower performance implementations that are unable to immediately calculate the acknowledgment. See Figure 5–34.

Direct link setup (DLS)

Direct link refers to the ability to exchange data directly between two STAs in the network without traversing the AP. The legacy IEEE 802.11 MAC specifies that STAs may only communicate with APs. When a STA sends a frame to another STA in the same logical network, the frame must be relayed through an AP. This approach ensures that communication is possible between all STAs, even if they are out of range to each other. But this approach reduces the available bandwidth for STA-to-STA communication by possibly more than one half. DLS in IEEE P802.11e provides a mechanism to allow direct STA-to-STA communication in the case where the STAs are in range of each other.

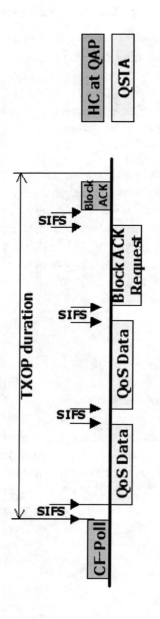

Figure 5–33: Overview of immediate block acknowledgment mechanism

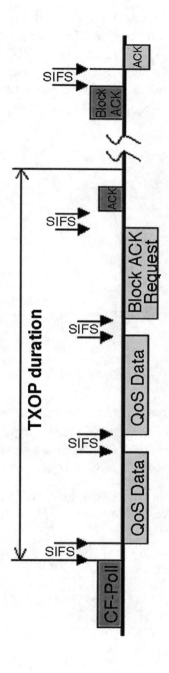

Figure 5–34: Overview of delayed block acknowledgment mechanism

The normal setup process for a direct link is shown in Figure 5–35. The STA wishing to initiate a direct link with another STA sends a DLS Request Action frame to the QAP. The QAP relays the request to the other STA, which responds with a DLS Response Action frame with a status of success. The QAP relays the response back to the STA that initiated the request. At that point direct communication may commence between the two STAs.

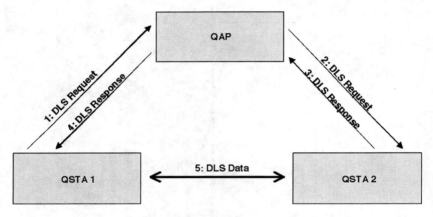

Figure 5–35:   Overview of DLS

There are several cases where the negotiation can fail. If the QAP does not allow DLS in the BSS or if the requested STA is not present in the BSS, the QAP may respond with a status of not allowed or not present to the first request. A STA may also respond with a status of refused to a DLS request.

A DLS teardown message can be used to explicitly end the DLS session. STAs also maintain a DLS inactivity timer that will time-out unused DLS connections.

Automatic power save delivery (APSD)

APSD is an enhancement of the existing IEEE 802.11 power save mechanisms. APSD allows a STA to set up a schedule for delivery of frames, based on a repeating pattern of a specified number of beacon periods.

When APSD has been enabled, the AP will buffer the APSD STA's frames for the number of beacon periods specified in the APSD setup. The time offset within the beacon period can be specified and allow a number of STAs to wake up at different times during one beacon period to receive their traffic.

APSD is especially useful for battery-operated devices that must turn off their radio entirely for the majority of the time, but still maintain reasonably low latency response to data sent from the QAP. The time offset parameter allows a larger number of devices to be supported by shifting the scheduled awakening time of different devices.

APSD support is indicated in the QAP's capability information. APSD operation is invoked by a STA by establishing a TSPEC with the APSD flag set. The QAP acknowledges APSD with a schedule element with the APSD flag set.

In addition to the scheduled APSD described above, an unscheduled APSD mode is available for QSTAs that are using EDCA access. Unscheduled service periods are initiated when the QSTA sends a QoS-Data/Null frame to the QAP and end when the QAP sends a Qos-Data/Null frame with the end of service period (EOSP) bit set.

IEEE P802.11e AS PART OF COMPLETE QoS IMPLEMENTATION

Scheduling and admission control

The AP is the focal point of a wireless network, typically operating at the interface between a wired network and the wireless medium with relatively limited bandwidth. The QAP is responsible for implementing scheduling and admission control to provide the QoS agreements negotiated with the wireless STAs. Scheduling and admission control are also required at STAs when the

STA supports multiple applications or traffic streams requiring different QoS parameters.

IEEE P802.11e does not specify how an HCCA scheduler must work, and there are many possible implementation techniques. IEEE P802.11e does specify the required signaling that is used over the air to convey the information needed by the scheduler to do its job.

The TSPEC is the primary mechanism for communication of QoS parameters. QSTAs send TSPEC requests to the QAP in the form of an ADDTS Action frames. The QSTA must request TSPEC for both upstream (from QSTA to QAP) and downstream (QAP to QSTA) flows. The QAP evaluates whether available resources exist to meet the requested TSPEC. The QAP can respond by offering the QSTA an alternate TSPEC (perhaps with lower performance QoS parameters), or it may deny the TSPEC request entirely.

Once a TSPEC has been established, it may be used for data transfer, and the QAP will meet the TSPECs QoS parameters to the extent possible. A TSPEC may be deleted by the QSTA or the QAP. The QAP may unilaterally delete a QSTA's TSPEC if there are changes in the channel condition that reduce available bandwidth or if higher priority TSPECs are requesting admittance.

TSPECs are generally created and destroyed based on requests from higher layer management entities. TSPECs would be deleted when the application using the QoS service has completed. Finally, a TSPEC will time out if corresponding traffic does not take place within the timeout defined during the setup.

In addition to the information contained in TSPECs, the scheduler can use information in the Queue Size subfield in the QoS Control field to help decide how to schedule TXOPs. The Queue Size subfield is present in QoS data frames sent by non-AP QSTAs.

Adapting to varying wireless channel conditions

The WLAN channel environment is challenging: Operation takes place in unlicensed spectrum where interference from other devices is commonplace, and the channel propagation properties can vary widely with movement of the wireless devices or objects in their vicinity. This variability makes the job of maintaining QoS exceptionally difficult.

The QAP is expected to grant TSPECs with specific QoS parameters appropriate to current channel conditions and to meet the request of a particular traffic stream or class to the best extent possible. At the same time, the PHY bit rates may have to be adjusted to account for changing channel conditions with any QSTA that associated. The QAP will have to adjust the scheduling and bandwidth allocations to compensate, and such adjustments may impact the QoS delivered to one STA when the conditions of the connection to another STA with a higher priority stream become degraded.

Interface to higher layers

The QoS mechanisms specified in IEEE P802.11e are intended to provide a general framework that is able to support present and future application requirements. Higher layers interface to the IEEE P802.11e MAC through the SAP, an abstract interface where data frames conceptually enter or leave the MAC sub-layer.

When bridging from other IEEE 802 MACs, IEEE P802.11e uses IEEE 802.1D priority tags associated with data frames at the MAC SAP. The UPs represented by IEEE 802.1D can range from 0 to 7.

These UPs are mapped to the traffic identifiers (TIDs) in the QoS Control field of QoS data frames transmitted and received on the wireless medium. The TID field is 4 bits and can represent values from 0 to 15. TID values from 8 to 15 represent traffic streams and are associated with corresponding TSPECs within the MAC.

The recommended priority ordering of IEEE 802.1D UP fields is not sequential from 0 to 7. Table 5–19 shows the mapping and designations of the priorities.

Table 5–19: Mapping between IEEE 802.1D UPs and EDCA ACs

UP -Same as IEEE 802.1D UP)	IEEE 802.1D Designation	AC	IEEE P802.11e Designation (Informative)
1	BK	1	Background
2	-	1	Background
0	BE	0	Best Effort
3	EE	2	Video
4	CL	2	Video
5	VI	2	Video
6	VO	3	Voice
7	NC	3	Voice

CONCLUSION

The completion of IEEE P802.11e will enable many new uses for WLAN. To date, the lack of QoS capabilities has held back many consumer and enterprise applications. Video and multimedia transport, as well as voice-over IP telephony, will become practical over WLAN, and new products will be introduced to take advantage of the QoS capability.

IEEE P802.11e provides two basic mechanisms to allow implementers to best address their application. EDCA provides simple prioritized access using contention-based probabilistic medium access. Differentiation of priority levels is achieved when averaged over time. HCCA enables parameterized QoS and supports multiple streams with individually tailored QoS requirements. Polled access from privileged HCs allows more predictable scheduling and reduced contention for higher efficiency. These two access mechanisms, combined with standardized signaling and a management interface to higher layers, provide the framework necessary for interoperable implementations of QoS for IEEE 802.11 WLANs.

Chapter 6 IEEE 802.11h dynamic frequency selection (DFS) and transmit power control (TPC)

In 1999, the European Union modified its regulations that applied to the radio band used by IEEE 802.11a. In 2003, the World Radiocommunication Conference (WRC 2003) adopted the harmonization and globalization of the 5 GHz frequency band from 5.150 GHz to 5.725 GHz confirming the use of IEEE 802.11a provided the new regulations are implemented. The new regulations required that any wireless device or system operating in the 5 GHz band must do four new things. First, the device must be able to detect the presence of radar operations. Second, the device or system must be able to avoid interfering with radar operations. Third, the system must be able to uniformly spread its operation across all the channels that may be used in the band. Fourth, the system must be able to minimize the overall power output of the system. These changes to the regulations were made to protect existing civil and military radar that already operates in the band, as well as to minimize the "hotspot" that might show up in urban areas in radar images of earth resource satellite systems, which also use this band.

To ensure that IEEE 802.11 WLANs operated in a consistent fashion when implementing systems to the new regulations, the IEEE 802.11 Task Group h (TGh) was chartered to write an amendment to the IEEE 802.11 standard. The amendment was ratified by the IEEE in July 2003. IEEE 802.11h defines mechanisms that can be used to measure radio channels for presence of radar, IEEE 802.11a activity, OFDM activity, and other radio activity. These mechanisms can be used individually by an IEEE 802.11 device, or they can be requested and coordinated by other IEEE 802.11 devices, such as APs. IEEE 802.11h also provides mechanisms to allow a BSS to move from one channel to another in a coordinated way when it is necessary to change the channel of the BSS while avoiding a radar signal.

Users of the 5 GHz frequency

Before 1999, much of the spectrum at 5 GHz for operation by wireless devices in Europe and in North America had been reserved for primary users, such as radar, satellite, and aeronautical and maritime navigation. Radio local area networks (RLANs) and IEEE 802.11a devices are considered secondary users of the frequency band. (See Figure 6–1.) In 2003, the FCC adopted a notice of proposed rulemaking allocating an additional 255 MHz of spectrum to the middle U-NII band ranging from 5.470 GHz to 5.725 GHz for users operating in the United States. Today the 5 GHz frequency band provides 23 channels of contiguous 555 MHz of spectrum for use by IEEE 802.11a devices, all of which are constrained to specific transmit power levels for specific blocks of spectrum. IEEE 802.11h addresses how IEEE 802.11a Wi-Fi STAs and APs perform the necessary procedures for operating in this frequency band. This chapter will describe the new capabilities and mechanisms introduced with IEEE 802.11h.

New parts to the IEEE 802.11 protocol

IEEE 802.11h added a number of information elements and a new management frame subtype, i.e., Action frame, to the IEEE 802.11 protocol. The Action frame implements two new request/response exchanges. The Action frame is also a generic frame type that can be utilized for many purposes beyond the uses described in IEEE 802.11h. The new information elements are used in the Action frame as well as in the Beacon, Probe Response, Association Request, Association Response, Reassociation Request, and Reassociation Response frames.

The Action frame is a general purpose container for management information. The format of the frame body is shown in Figure 6–2.

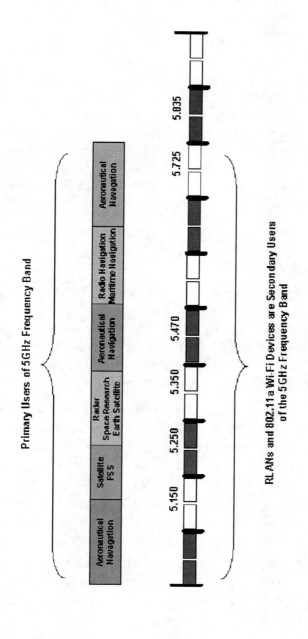

Figure 6–1: The 5 GHz frequency band

Order	Information
1	Action

Figure 6–2:   Action frame body format

The Action frame is a typical management frame. In the Frame Control field of its MAC header, it has a Type subfield equal to zero and a Subtype subfield equal to 0b1101, a combination defined as reserved in the IEEE 802.11 base standard. Otherwise, the format of the Action frame's MAC header is identical to other management frames. The frame body of the Action frame contains a single, variable-length field, i.e., Action. The Action field comprises two subfields, i.e., Category and Action Details, as shown in Figure 6–3.

Category	Action Details
1 octet	Variable octets

Figure 6–3:   Category and Action Details subfields

In addition, bit 8 in the 2-octet Capability Information field is assigned to indicate STA support for IEEE 802.11h spectrum management. If a STA supports IEEE 802.11h spectrum management, then bit 8 is set to a true logic 1; otherwise, the bit is set to logic 0. See Figure 6–4.

The Category subfield is a single octet and identifies the major function of a particular Action frame. IEEE 802.11h defines the value of zero for the Category subfield to indicate that the Action frame's major function is spectrum management. Values of 1-127 and 129-255 are reserved in IEEE 802.11h. The Category subfield is used by the recipient of an Action frame to indicate an error by setting the MSB of the Category subfield value and returning the Action frame to the sender with the remainder of the frame unchanged.

The Action Details subfield is of variable length and contains the information necessary to carry out the action being requested of the recipient by the sender. For IEEE 802.11h, this subfield contains an information element with the specific action required to be accomplished.

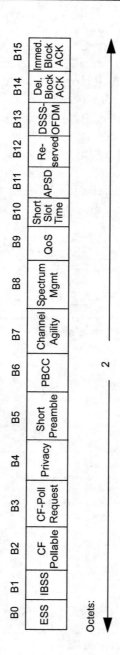

Figure 6–4: Capability Information field format

One confusing aspect of IEEE 802.11h is the use of terms such as "TPC Request" frame, "TPC Report" frame, and other similar names for apparently new frame types. These names actually refer to an Action frame with an information element of the named type in the Action Details subfield. Thus, a "TPC Request" frame is an Action frame with the TPC Request information element present in the Action Details subfield. Similarly, a "TPC Report" frame is an Action frame with the TPC Report information element present in the Action Details subfield. Other "frame" types mentioned in IEEE 802.11h are "Channel Switch Announcement," "Measurement Request," and "Measurement Report" frames. All of these names refer to Action frames with the corresponding information element in the Action Details subfield of the frame.

In addition to the Action frame, IEEE 802.11h defines 12 new information elements, one octet each as shown in Table 6–1.

Table 6–1:   IEEE 802.11h information elements

Item	Information element	New element ID
2	Power Constraint	32
3	Power Capability	33
4	TPC Request	34
5	TPC Report	35
6	Supported Channels	36
7	Channel Switch Announcement	37
8	Measurement Request	38
9	Measurement Report	39
10	Quiet	40
11	IBSS DFS	41
12	ERP Information	42
14	Extended Supported Rates	50

Transmit Power Control (TPC)

TPC is one of the new features added to IEEE 802.11 WLANs to help reduce the amount of interference across all channels and with satellite services occupying the 5 GHz frequency band. TPC is specified to operate with existing legacy IEEE 802.11 STAs and APs. The ability to associate and operate between the STA and APs is based on the STA's transmit radio frequency (RF) power capabilities. STAs and APs must have the capability built into the radio (i.e., PHY) implementation to vary the RF transmit power over a range to satisfy the requirements specified by global regulatory bodies. A STA's capability to adapt to a specific RF transmit power level is based on path loss and RF link margin calculations. The following paragraphs describe the specific functions that are necessary for TPC.

The Power Constraint information element consists of three fields, each 1 octet long: Element ID, Length, and Local Power Constraint. These information elements are used by the STA to calculate the maximum RF transmit power measured in decibels for the current operating frequency channel as shown in Figure 6–5. The Length field is always set to logic 1. The Local Power Constraint field contains a power level offset value, which is stored as an unsigned integer in decibels. The Local Power Constraint is used as a variable to normalize the STA's RF transmit power for the selected operating frequency channel. The local maximum transmit power allowed a STA is defined as the difference between the maximum transmit power allowed for a given channel as specified for a country and the value stored in the Local Power Constraint field. (See the equation below.) The Power Constraint information element is stored as MIB and is used during Beacon frames and Probe Request frames by the STA when communicating to APs.

Element ID	Length	Local Power Constraint

Figure 6–5:   Power Constraint information element

Local maximum power allowed is defined as follows:

$$\text{RF TxPower}_{Max} \text{ dB} = \text{Country RF Channel TxPower}_{Max} - \text{Local Power Constraint}$$

Each STA must have keep a record of the transmit power capability when requested by APs during Association Request frames and Reassociation Request frames. The Power Capability information element consists of four fields, each 1 octet long, as shown in Figure 6–6.

Element ID	Length	Minimum Transmit Power Capability	Maximum Transmit Power Capability

Figure 6–6:   Power Capability information element

The Minimum Transmit Power Capability field and Maximum Transmit Power Capability field contain the value of the minimum and maximum transmit RF power level, respectfully, for the STA operating on the current frequency channel. The value in the fields is a signed integer in decibel increments of 1 mW with a tolerance of $\pm$ 5 dB.

When TPC is enabled in an IEEE 802.11 WLAN, the AP requests the STA to report the transmit power level and link margin information of the PHY. This information is reported by using the TPC Report information element. The TPC Report information element is 2 octets long with two fields, i.e., Element ID and Length, where the Length field is set to logic 1. Through a sequential set of TPC Request information element responses, the TPC Report information element for the STA as shown in Figure 6–7 is sent back to the AP.

Element ID	Length	Transmit Power	Link Margin

Figure 6–7:   TPC Report information element

During the TPC report request, the Transmit Power field is set to the transmit RF power in the PHY used to transmit the frame containing the report. The field is a signed integer of decibels, with a tolerance of $\pm$ 5 dB. The tolerance is defined as the difference between the reported transmit RF value and the actual equivalent isotropically radiated power (EIRP) of the STA as measured over 1500 octets.

The Link Margin field is a signed integer value in decibels, containing the link margin at the time for the data rate of the TPC Request information element received. Link margin is calculated using free space path loss theory, and the value is dependent upon the implementation of the PHY and environment of the operating frequency channel.

The Supported Channels information element as shown in Figure 6–8, is 4 octets long and contains the number of 5 GHz frequency channels supported by the STA.

Element ID	Length	First Channel	Number of Channels

Figure 6–8:   Supported Channels information element

The value in the Length field is defined by the First Channel field followed by the number of channel pairs. The First Channel field is set to the first channel as specified for the subbands of the supported frequency channels supported. The Support Channels information element is transmitted from the STA to the AP through the use of Association Request frames.

TPC operation

The operation of TPC is based on a sequence of Beacon frames and Probe Response frames along with Association Request/Response frames and Reassociation Request/Response frames between the STA and APs. IEEE 802.11h was designed to support legacy IEEE 802.11 WLANs. APs may allow association of legacy Wi-Fi STAs that do not support TPC functionality providing they support the country-specific regulatory requirements. If an AP receives a STA request to associate using the Power Capability information element, the STA must respond by reporting the minimum and maximum transmit power level supported for the frequency channel in use using a Power Capability information element. However, if an AP decides that the STA's minimum and maximum levels are not sufficient to join the WLAN, then the AP may reject the STA from joining. For some regulatory domains, the Country information element contains the value of the local maximum transmit power for operation in the 5 GHz band. The value for the local maximum transmit power level may change on the BSS life cycle. Once the optimal transmit power levels are established for an AP, the minimum and maximum values are very rarely changed for the network. It is required that all APs and ad-hoc STAs supporting this functionality make it known that this feature is supported. This announcement is accomplished through the use of the Country information elements contained in the Beacon frames and Probe Response frames.

SPECTRUM MANAGEMENT

Spectrum management is a key feature specified in IEEE 802.11h, which provides several mechanisms to avoid co-channel operation of WLAN STAs with civil and military radars operating in any or all IEEE 802.11a channels across the 5 GHz frequency band. This feature is referred to as *dynamic frequency selection* (DFS).

DFS is accomplished when the AP requests STAs to make channel power measurements of the channel of interest over finite periods of time and reports the results so that the AP can command any STA in the wireless network to take the appropriate actions if radars are present. The following paragraphs describe the several measurement requests and reports for spectrum management.

Using the Measurement Request information element, IEEE 802.11h specifies three types of measurement requests as part of the spectrum management between STAs and APs. See Table 6–2.

Table 6–2:   Measurement types

Name	Measurement type	STA response to AP
Basic request	0	Mandatory
Clear channel assessment (CCA) request	1	Optional
Receive power indication (RPI) histogram request	2	Optional
Reserved	3–255	

In addition to having a Measurement Type field, the Measurement Request information element has a Measurement Request field, which has subfields based on the type of measurement requested. The basic request consists of three subfields: Channel Number (1 octet), Measurement Start Time (8 octets), Measurement Duration (2 octets). See Figure 6–9. A STA must be capable of generating a basic report to the associated AP.

Channel Number	Measurement Start Time	Measurement Duration

Figure 6–9:   Measurement Request field format for a basic request

The Channel Number subfield corresponds to the frequency channel for which the measurement is obtained. The measurement will be taken over a defined period by a start time and duration specified in the Measurement Start Time and Measurement Duration subfields, respectively. The start time is set to the timing synchronization function (TSF) timer ($\pm$ 32 s). The timer will start immediately if the value of the Measurement Start Time subfield is set to zero. The value in the Measurement Duration subfield is an integer measured in TUs.

The CCA request is a STA response to a CCA request by the AP. The STA response is optional for this report. The CCA request consists of three subfields: Channel Number (1 octet), Measurement Start Time (8 octets), Measurement Duration (2 octets). See Figure 6–10.

Channel Number	Measurement Start Time	Measurement Duration

**Figure 6–10:   Measurement Request field format
for a CCA request**

The measurement and field structure for the CCA request is analogous to the basic request and uses the same TSF timer functions.

The RPI histogram request is a STA response to a RPI request by the AP. The STA response is optional for this report. The RPI histogram request consists of three subfields Channel Number (1 octet), Measurement Start Time (8 octets), Measurement Duration (2 octets). See Figure 6–11.

Channel Number	Measurement Start Time	Measurement Duration

**Figure 6–11: Measurement Request field format
for an RPI histogram request**

The measurement and field structure for the RPI histogram request is analogous to the basic request and uses the same TSF timer functions.

With the Measurement Report information element, three types of measurement reports are used in response to the measurement types for communication between STAs and APs. See Table 6–3.

Table 6–3: Measurement reports

Name	Measurement type	STA report to AP
Basic report	0	Mandatory
CCA report	1	Optional
RPI histogram report	2	Optional
Reserved	3–255	

In addition to having a Measurement Type field, the Measurement Report information element has a Measurement Report field, which has subfields based on the type of measurement being reported. The basic report consists of four subfields: Channel Number (1 octet), Measurement Start Time (8 octets), Measurement Stop Time (2 octets), and Map (1 octet). The operation for the basic report is analogous to the basic request. The MAP subfield consists of 8 bits, which identify several status flags for reporting. See Figure 6–12.

BSS	OFDM Preamble	Unidentified Signal	Radar	Unmeasured	Reserved
B0	B1	B2	B3	B4	B5-B7

Figure 6–12:   Map subfield format

The following bits are set accordingly if the associated statement is true:

B0: Set to logic 1, if a single valid MPDU was received

B1: Set to logic 1, if a valid short IEEE 802.11a OFDM short symbol was detected

B2: Set to logic 1, if energy is present, but an OFDM short symbol or radar was not detected

B3: Set to logic 1, if a radar was detected during the measurement period

B4: Set to logic 1, if this channel was not measured (When set to logic 1, all other bits are set to logic 0.)

When a basic request is sent to the STA from the AP, the STA begins measurement and the appropriate bits are set in the MAP subfield as part of the basic report.

The RPI histogram report starts with three subfields: Channel Number (1 octet), Measurement Start Time (8 octets), Measurement Duration (2 octets), as shown in Figure 6–13. Those subfields are followed by eight subsequent subfields, called RPI Density (1 octet each), as shown in Figure 6–14.

Channel Number	Measurement Start Time	Measurement Duration

Figure 6–13:   First three subfields of a RPI histogram report

RPI 0	RPI 1	RPI 2	RPI 3	RPI 4	RPI 5	RPI 6	RPI 7
Density	Density	Density	Density	Density	Density	Density	Density

Figure 6–14:   Subsequent RPI density subfields of a RPI

Once the RPI histogram request is received, the STA will form a histogram by starting to measure the receiver channel power at the antenna port for the duration specified in the Measurement Duration subfield. The receive power levels are converted to one of the RPI density levels listed in Table 6–4, using the following equation:

$$\text{RPILevel} = (255 \times (\text{Duration}_{\mu s} / (1024 \times \text{Measurement duration})))$$

Table 6–4:   RPI levels

RPI	Receive power at the antenna port (0 dBm)
0	Power ≤ -87
1	$-87 <$ Power ≤ -82
2	$-82 <$ Power ≤ -77
3	$-77 <$ Power ≤ -72
4	$-72 <$ Power ≤ -67
5	$-67 <$ Power ≤ -62
6	$-62 <$ Power ≤ -57
7	$-57 <$ Power

The RPI histogram report is a very useful tool as it provides information about neighboring STAs in terms of interference and reduces the probability of false alarms for detecting radars.

The Quiet information element consists of six fields: Element ID (1 octet), Length (1 octet), Quiet Count (1 octet), Quiet Period (1 octet), Quiet Duration (2 octets), and Quiet Offset (2 octets). See Figure 6–15.

Element ID	Length	Quiet Count	Quiet Period	Quiet Duration	Quiet Offset

Figure 6–15:   Quiet information element

The Quiet information element specifies a period of time by which the STA stops at all transmission of the packets over the RF medium. The Quiet information element is passed to the STA from the AP using Beacon frames and Probe Request frames. This feature is used in conjunction with measuring and gathering channel characteristics, such as detecting radars, reporting receive power RPI. It allows for gathering measurements about the channel that are more reliable statistically without worrying about interference from neighboring IEEE 802.11h-supported ad-hoc and networked STAs.

DETECTION OF RADAR OPERATION

IEEE 802.11h specifies that a STA must discontinue the transmit operation of RF energy in the frequency channel if radars are present. Listed in Table 6–5 are three radar signal types specified in ETSI EN 301 893 [v.1.2.3 (2003-08) Annex D] as referenced in IEEE 802.11h. This list is just a sample what types of radars may occupy the 5 GHz spectrum. The pulse widths range from 1 μs to 2 μs and repetition rates ranging from 330 pps to 1800 pps for a duration ranging in milliseconds.

Table 6–5:   Radar types specified by ETSI

Radar signal type	Pulse width (μs)	Repetition rate (pps)	Burst length (ms)	Minimum number of pulses detected
1	1	700	26	Detect 18 pulses
2	1	1800	5	Detect 10 pulses
3	2	330	210	Detect 70 pulses

Implementations for detecting radar signals are not specified in IEEE 802.11h. It is the responsibility of the product designer to implement such detection mechanisms. However, Figure 6–16 illustrates an example of a typical implementation splitting the detection and monitoring functions between the PHY and MAC for radar detection. The basic method of detection is based on monitoring the RF energy [i.e., receive signal strength indicator (RSSI)] in the receiver and using a threshold level detector to trigger events. From the example shown in Figure 6–16, if an IEEE 802.11a OFDM signal is not demodulated, then the signal energy is assumed to be one of the following:

a) Noise in the frequency channel

b) An unknown interference source

c) Other users of the frequency band, such as radars.

If the signal energy is believed to be a radar, it is declared valid only after detecting the minimum number of pulses for the radar type as illustrated in Table 6–5. Upon detecting radars, the STA will stop all data transmissions and set the appropriate status bits in the basic report as previously described.

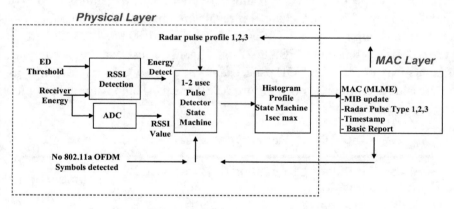

Figure 6–16:   Radar pulse detection

Chapter 7 IEEE 802.11d international operation

With the publishing of IEEE 802.11b in 1999, the IEEE 802.11 Working Group was beginning to see the acceptance of products based on IEEE 802.11b. These products were exciting a demand that had not occurred with the IEEE 802.11 base standard. There was a problem with IEEE 802.11b, however. It was defined to operate only in the United States, Canada, Japan, France, Spain, and the portion of Europe operating under ETSI regulations. Anywhere outside of these six locations, or *regulatory domains* as they are defined in the standard, a device could not be called compliant with the IEEE 802.11 standard. To address the problem of defining compliant operation outside of the original six regulatory domains defined in the standard, IEEE 802.11 Task Group d (TGd) was formed.

The official name of IEEE 802.11d is "Operation in Additional Regulatory Domains." The purpose of the amendment is to allow operation of 802.11 devices on every square meter of planet Earth. To this end, TGd examined a number of alternatives that would meet this purpose. At the beginning, it was expected that TGd would incorporate specific regulatory information into the standard. However, that solution would have been quite onerous to implement and quickly become obsolete, as governments changed local regulations.

The solution chosen by TGd was to add a management protocol that would announce certain regulatory and location information. This protocol allows mobile devices to identify where they were operating; to determine whether they were allowed to operate in that location; and, if operation was allowed, to configure their radio to comply with local regulations. This protocol extended the information in the Beacon, Probe Request, and Probe Response frames with new information elements providing information on location, regulatory requirements, specific configuration for FH PHYs, and a mechanism in the Probe Request frame for a STA to request specific information elements to be returned in the Probe Response frame.

NEW ROAMING REQUIREMENTS

With the inclusion of the Country information element, IEEE 802.11d makes it possible for compliant equipment to be dynamically configured to operate in new regulatory domains. Colloquially, this concept is *international roaming*. The Country information element provides a country code and regulatory information about the radio band and channels in use. See Figure 7–1.

Element ID	Length
Country String (Octets 1,2)	
Country String (Octet 3)	First Channel Number
Number of Channels	Maximum Transmit Power Level
• • •	
First Channel Number	Number of Channels
Maximum Transmit Power Level	Pad (if needed)

Figure 7–1:   Country information element

What the Country information element *does not* do is make all IEEE 802.11 equipment automatically capable of operating all over the world. The reason is that most radio equipment must receive certification by the local regulatory agency before it is allowed to operate within the boundaries controlled by that regulatory agency. This requirement is true even of unlicensed devices, such as devices described by the IEEE 802.11 standard. Even though the operator of IEEE 802.11 equipment does not need to hold a license to operate the device, the manufacturer must ensure to the satisfaction of the regulatory agency that the device meets all the regulatory requirements in order to be granted the certification that is necessary to allow the equipment to be legally sold and operated. The manufacturer of IEEE 802.11 equipment is not relieved of the responsibility of preventing the operation of their devices in regulatory domains in which it has not obtained the required certifications.

What the Country information element *does* do is allow a manufacturer to build a radio that can be configured to operate in a large number of different regulatory domains and to update the radio's configuration information with new configurations as the radio is certified in more regulatory domains. This step can be accomplished by downloading updated firmware for the radio, for example. Only when the manufacturer provides specific configuration information for a regulatory domain should the device be allowed to operate in that regulatory domain. One implication of this requirement is that the device must include a list of regulatory domains in which it is allowed to operate. Upon encountering a country code indicating a regulatory domain that is not on its list, the radio must not transmit.

IEEE 802.11 STAs compliant to IEEE 802.11d begin operation with passive scans to determine whether there are any Beacon frames that can be received that contain a Country information element. If the STA receives a Country information element, it must compare the country code in the information element against the list of regulatory domains in which it is allowed to operate. If a match is found, the STA must configure its radio according to internal information for the regulatory domain so that it does not operate outside of the channel and transmit power information included in the Country information element. If there is not a match for the regulatory domain indicated in the Country information element, the STA must not transmit, as the STA does not have a configuration programmed that ensures operation compliant with local regulations.

If the STA receives Beacon frames, but none that contain Country information elements, or does not receive any Beacon frames, the STA is required by the standard not to begin any transmissions. This requirement is much more restrictive than for STAs produced prior to IEEE 802.11d. The reasoning is that a STA implementing IEEE 802.11d is capable of recognizing the regulatory domains in which it is allowed to operate and must not allow operation outside of those regulatory domains. In the absence of any information about the regulatory domain, the STA must assume that it is not in a regulatory domain where it is certified to operate.

Country information element

The Country information element is a compact structure to convey information identifying the regulatory domain and the bands and channels in use according to the regulations in force. The format of the Country information element is shown in Figure 7–1.

The structure of the Country information element contains the standard Element ID and Length fields of all IEEE 802.11 information elements. The element ID of the Country information element is 7. The value of the Length field is equal to the number of bytes following the Length field. For the Country information element, the value of the Length field is always even. The Country information element is padded with 0 or 1 byte so that the length of the information element is always an even number of bytes.

The Country String field is 3 bytes long. The value of the field is three ASCII characters. The first two characters are the two-character identifier for the country, obtained from the alpha-2 column of the table titled "Codes for the representation of names of countries and their subdivisions" in clause 9 of ISO/IEC 3166-1. The third character may be either an ASCII "I" or "O" to denote indoor or outdoor operation where such regulations differ or and ASCII space character (" ") where common regulations define both indoor and outdoor operation.

Following the Country String field is one or more triplets defining the channelization and maximum transmit power for each radio band available for use in the regulatory domain. Each triplet contains the following fields: First Channel Number, Number of Channels, and Maximum Transmit Power Level. The value of the First Channel Number field is an integer, identifying a channel number from the PHY specification for the radio on which the Beacon frame containing the Country information element was received. This first channel number identifies the lowest channel number to which the radio can be configured to operate in the band described by the triplet. The first channel number is not a value that can be selected by the manufacturer. It is defined by the regulations for the band and the PHY definition. For example,

IEEE 802.11b defines 14 channels between 2400 and 2500 MHz. For the "US" regulatory domain, the first channel must be 1, identifying the first channel allowed to be used by the regulations. A manufacturer may not change this value to restrict the operation of the STAs receiving the Country information element.

The value of the Number of Channels field is also an integer, identifying the total number of legal channels available in the band described by the triplet. This field identifies only the number of *legal* channels available in the band, as defined in the PHY specification for the radio on which the Beacon frame containing the Country information element was received. It is often thought, incorrectly, that this field defines the number of *usable* channels in the band. An example of the difference between the number of legal channels in a band and the number of usable channels in a band would be the 2.4 GHz ISM band in the United States for DSSS operation, where the number of legal channels is 11, but the number of usable channels (where usable is maximum number of nonoverlapping channels) is 3. Again, this field must carry the information that pertains to the regulations and the PHY in use. It may not be changed by a manufacturer to further restrict the operation of the STAs receiving the Country information element.

The Maximum Transmit Power Level field is an integer, identifying the maximum transmit power (in dBm) allowable in the band. The interpretation of this field varies according to the measurement method specified in the particular regulations of the regulatory domain. In some regulatory domains, this value may indicate the maximum transmit power measured at the antenna connector into a standard load. In other regulatory domains, this value may indicate the effective isotropic radiated power (EIRP) of the radio and its attached antenna. A manufacturer must understand the measurement requirements of a regulatory domain in order to interpret this field properly.

The Country information element may include more than one triplet, where each triplet describes a single band covered by common regulations. When more than one triplet is present in the information element, the triplets must appear in the information element in order of increasing first channel number. An example of such regulations is the US U-NII band, where three subbands are defined. A Country information element describing the complete U-NII band would include three triplets, each describing one of the three 100-MHz-wide subbands.

For the IEEE 802.11a, IEEE 802.11b, and IEEE 802.11g PHYs, the Country information element is sufficient to convey all the information necessary for international roaming. For FH PHYs, however, additional information is necessary.

International roaming with FH PHYs

In addition to the information about the radio bands carried in the Country information element, FH PHYs require additional information in order to synchronize the operation of the radio in the STA with the operation of the radio in the AP. STAs using FH PHYs must be able to identify the exact hopping sequence that is being used by an AP. In the IEEE 802.11 base standard, a number of hopping patterns and sets are defined for the 2.4 GHz band, and they could be reused in any radio band of equal or greater bandwidth. However, that number may not be sufficient for all bands. In order to address this possibility, IEEE 802.11d incorporates two new methods to define hopping patterns. The first is an algorithmic construction. The second is based on tables. Only one of the two methods may be used by an AP to convey information to a STA about its hopping pattern.

FH Parameters information element

The FH Parameters information element provides the information required to calculate a pseudo-random hopping pattern from only two values. The format of the FH Parameters information element is shown in Figure 7–2.

Element ID	Length
Prime Radix	Number of Channels

Figure 7–2:   FH Parameters information element

The Element ID of this information element is 8. The length of the information element is 4 bytes. The value of the Length field of the information element is 2. This information element contains two fields: Prime Radix and Number of Channels. Each field is 1 byte long. The values from the Prime Radix and Number of Channels fields are inserted into a set of equations to calculate a pseudo-random hopping pattern. The equations generate a set of pseudo-random sequences, called *hyperbolic congruence codes* and *extended hyperbolic congruence codes*. These codes are then offset by the first channel number in a band to produce the hopping patterns. Because the equations produce a set of congruence codes, the FH Parameter Set information element specifies which code sequences are to be used to determine the particular hopping pattern in use by the AP.

Hopping Pattern Table information element

The Hopping Pattern Table information element is the second method provided by IEEE 802.11d to determine the hopping pattern used by the AP. This information element contains the following fields: Flag, Number of Sets, Modulus, Offset, and, optionally, a Random Table. The format of this information element is shown in Figure 7–3.

Element ID	Length
Flag	Number of Sets
Modulus	Offset
Random Table (Octets 1,2)	
• • •	
Random Table (Octets n-1, n)	

Figure 7–3:   Hopping Pattern Table information element

The Element ID of this information element is 9. When this information element is present, the value of the Flag field determines whether the *hop index* method or *random table* method is used to calculate the hopping pattern. When the Flag field is zero, the hop index method is to be used, and a Random Table field is not present in the information element. When the Flag field is one, the Random Table field is present in the information element, and it is to be used to calculate the hopping pattern.

REQUEST INFORMATION ELEMENT AND ITS PROTOCOL

During the development of IEEE 802.11d, the participants in TGd became concerned at the amount of information that was being put into the Beacon frame. The additional information was from IEEE 802.11d and from other work in development at that time as well. The reason for the concern was that anything placed in the Beacon frame becomes fixed overhead for the WLAN, i.e., time consumed that cannot be used for communication by the associated STAs. To address this concern, IEEE 802.11d extended the functionality of the Probe Request and Probe Response frames through the inclusion of a new information element: the Request information element. The first use that IEEE 802.11d envisioned for this new information element and request-response protocol was to remove the need to carry the Frequency Hopping Table information element from the Beacon frame.

The Request information element is the initiator of a simple request-response protocol, carried by the Probe Request and Probe Response frames. The format of the Request information element is shown in Figure 7–4.

The Element ID of the Request information element is 10. Following the Length field is a list of information element IDs that are requested to be returned in the Probe Response frame that results from the reception of the Probe Request frame containing the Request information element. The element IDs must be included in the information element in monotonically increasing order. Any other order is considered to result in a malformed information element, and the responding STA may choose not to return any information elements following the last correctly ordered element ID in the Request information element.

Element ID	Length
Requested Element ID 1	Requested Element ID 2
. . .	
Requested Element ID n-1	Requested Element ID n

Figure 7–4:   Request information element

The request protocol is very simple. A STA includes a Request information element, with a list of element IDs of the information elements to be returned, in a Probe Request frame. The Probe Request frame is sent to either a broadcast address or to the individual address of another STA. To respond to a Probe Request frame, the STAs include in their Probe Response frames the information elements from the element ID list of the Probe Request frame that the STAs understood. The Probe Response frame is sent directly to the individual address of the STA that sent the Probe Request frame.

Chapter 8 IEEE 802.11F Inter Access Point Protocol (IAPP)

IEEE 802.11F defines a recommended practice for a protocol to be used between IEEE 802.11 APs: the Inter Access Point Protocol (IAPP). IAPP is to aid in the exchange of information between APs when a mobile STA roams from one AP to another. Because IEEE 802.11F describes a protocol above the data link layer, it is not an IEEE standard. This chapter provides a description of IAPP.

GOING BEYOND THE MAC

While there are a large number of WLANs that have only a single AP, such as a WLAN in a small office or in a home, many WLANs are built with dozens or hundreds of APs. The IEEE 802.11 standard defines the over-the-air MAC sublayer protocol between two IEEE 802.11 devices, such as a mobile STA and an AP. This definition includes the management protocol to make layer 2 roaming from one AP to another AP transparent to higher protocol layers. However, it is not sufficient to provide only the over-the-air protocol to enable transparent roaming. The AP is also connected to the DS, which is usually another IEEE 802 network with embedded IEEE 802.1D bridges. The layer 2 devices in the DS must also be informed of the new location of the mobile STA when it roams in order to direct the frames for that STA to the correct bridge ports to reach the AP where it is associated. This need has been dealt with by manufacturers in a proprietary fashion, since IEEE 802.11 began.

The goals of IEEE 802.11F are to provide a recommended way for APs to communicate with each other when a mobile STA roams between them, to describe a way for an AP to update the forwarding tables in IEEE 802.1™ MAC bridges that may be included in the IEEE 802.11 DS, to establish a format for APs to exchange context information about a STA that has roamed, and to enable the distribution of STA context information from one AP to neighboring APs. To meet these goals, IEEE 802.11F defines a new protocol, IAPP, and rules for its operation.

Some of the things that IEEE 802.11F does not define is what an AP should put into the context container to send to a peer AP or what format the information in the context container should have. This was purposely left for future standardization, which has not yet taken place. Some uses of the context container have been proposed, e.g., pushing IEEE 802.11i keying information to neighboring APs or carrying QoS information for IEEE P802.11e. Unfortunately, neither IEEE 802.11i. nor IEEE P802.11e define the use of the context container for those amendments. For this and other reasons, IEEE 802.11F is not seen as a protocol that is vital to the IEEE 802.11 industry. Because IEEE 802.11F is a trial use document, it must be reaffirmed within two years of its approval. If it is not reaffirmed, IEEE 802.11F will be withdrawn.

MORE ABOUT MOBILITY

IEEE 802.11 describes the protocol exchanges necessary for mobility on the wireless side of an AP. On the wired side of an AP (i.e., the DS), network configuration changes need to be made to ensure that the packets sent to the roaming mobile client arrive at the correct destination. At the simplest level, in other words, the bridges and switches that connect the APs together on the wired network must be triggered to update their forwarding tables and send frames with the MAC address of the mobile STA to the correct port on the bridge or switch so that those frames will reach the AP where the STA is currently associated.

Most bridges and switches are continuously in learning mode, and they update forwarding tables for each port on the bridge or switch whenever a new MAC address is seen on a port. While there is much more to this process than described here, for the purposes of IEEE 802.11 mobility, it is sufficient to say that a bridge or switch will learn that a new device is attached to a particular port when it receives a frame with a MAC address in the SA field that has not previously been seen on that port. (The SA field is second address field of an Ethernet frame; the first field is the DA.) Upon receiving a frame with a new MAC address, the bridge or switch will update the forwarding table for that port and add the new MAC address with an indication that it is attached to the port where the frame was received. If necessary, the bridge or switch will remove an entry for that same MAC address that might have been previously seen on a different port. Once the MAC address is in the forwarding table, the bridge or switch forwards to the port designated in the forwarding table all frames with the newly learned MAC address in the DA field received on any port other than the one on which the MAC address was learned.

IEEE 802.11i introduced a new wrinkle into the problem of updating bridges and switches after roaming from one AP to another. For IEEE 802.11i, connectivity between the STA and the DS is not provided immediately after association. When using IEEE 802.11i, the STA must complete the Authentication and Key Management Protocol (AKMP, see Chapter 4) and the 4-Way Handshake establishing the pairwise transient keys (PTKs, see Chapter 4) before the authenticator in the AP will allow frames to pass between the STA and the DS.

IEEE 802.11F has not been updated to change the sequence of events that IAPP follows when sending the layer 2 update frame immediately upon deciding to accept the association of a STA. To operate with IEEE 802.11i, the layer 2 update frame should not be sent by the AP until the authenticator allows frames to pass over the IEEE 802.1X controlled port between the STA and the DS.

Chapter 9   MAC management

IEEE 802.11 is the first LAN standard by the IEEE 802 Committee that includes significant management capabilities. The reason is that an IEEE 802.11 WLAN must deal with an environment that is measurably more complex than the environments of the wired LAN standards of IEEE 802. The largest challenge for the IEEE 802.11 WLAN is that the medium is not a wire. It is this simple fact that leads to all of the other obstacles that IEEE 802.11 must overcome in order to offer the same reliable service expected of an IEEE 802 LAN.

Because the media over which the IEEE 802.11 WLANs operate are not wires, the media are shared by other users that have no concept of data communication or sharing the media. An example of this type of user is the common microwave oven. The microwave oven operates in the 2.4 GHz ISM band because one excitation frequency of the water molecule lies in this band. The oven operates by transferring energy to the water molecules in food, thereby heating the food. Another user in this same band is the radio frequency ID (RFID) tag. RFID tags are usually small, cheap, unpowered devices that receive their power from a microwave beam and then return a unique ID. RFID tags are used to track retail inventory, identify rail cars, and do many other services. An unfortunate consequence of these devices sharing the band with WLANs is that some of this microwave energy leaks from the oven and is purposely broadcast for RFID tags. These leaks and broadcasts interfere with the operation of the WLAN.

WLANs other than IEEE 802.11 WLANs share the media. This existence is somewhat equivalent to attempting to run IEEE 802.3, IEEE 802.5™, IEEE 802.12™, and finer distributed data interference (FDDI) on the same twisted pair cable, simultaneously. These other WLAN users of the media are often uncoordinated with IEEE 802.11 and, in most cases, do not provide for any mechanism to share the media at all. Finally, there are other IEEE 802.11 WLANs sharing the media.

These other users of the media result in the first challenge for the IEEE 802.11 WLAN: an intermittent connection to the media and other STAs of the IEEE 802.11 WLAN. Much of the management specified in IEEE 802.11 is to deal with the intermittent nature of the media.

The second challenge for the IEEE 802.11 WLAN is openness, i.e., anyone can "connect" to the WLAN, simply by erecting the right kind of antenna. This openness leads to the need to identify the STAs connecting to the WLAN, in order to allow only authorized STAs to use the WLAN, and also leads to the need to protect the information sent over the WLAN from improper interception by eavesdroppers.

The third challenge for the IEEE 802.11 WLAN is mobility. Once the wires are removed from a LAN, the natural thing to do is to pick up the equipment connected to the LAN and move it around, taking it from an office to a conference room or to another building. Thus, IEEE 802.11 equipment is not always in the same place from one moment to the next. Even if the equipment were to remain in a fixed location, the nature of the wireless media may make it appear as if the equipment has moved. Dealing with mobility while making all of the expected LAN services available is a problem to be solved by MAC management.

The final challenge for the IEEE 802.11 WLAN is power management. Another consequence of doing away with wired media and enabling the equipment to be mobile is that much of the equipment will be run on batteries. Conserving the energy stored in the batteries to allow the equipment to operate for as long as possible must be built into the WLAN protocol and controlled by MAC management.

TOOLS AVAILABLE TO MEET THE CHALLENGES

The IEEE 802.11 standard defines a number of MAC management capabilities that are designed to meet the challenges of operating a reliable WLAN. These tools are authentication, association, address filtering, privacy, power management, and synchronization.

Authentication

Authentication provides a mechanism for one STA to prove its identity to another in the WLAN. The process of authentication is the exchange of questions, assertions, and results. An example of this exchange would be STA A asserting "I am STA A," and asking STA B "Who are you?" At this point, the process of authentication varies dependent on the particular algorithm in use. It may proceed with STA B saying "OK, prove you are STA A" and asserting "I am STA B." STA A would then offer some proof of its identity and request that same kind of proof from STA B. If the proofs exchanged were acceptable, each STA would then tell the other that its assertion of identity is believed.

Authentication can be used between any two STAs. However, it is most useful when used between a mobile STA and an AP in an infrastructure LAN. In this case, the AP is the point of entry for any mobile STA into the ESS and, possibly, into the wired LAN behind the ESS. Full proof of the identity of a mobile STA is necessary if the network is to be protected from unauthorized users. Authentication is also very useful between APs connected via a wireless DS, such as in a mesh network. IEEE 802.11i has made significant improvements in the authentication of IEEE 802.11 devices. See Chapter 4 for more information on IEEE 802.11i.

As it is defined in IEEE 802.11, there are two authentication algorithms available. The first algorithm, open system authentication, is not really an authentication algorithm at all. It is a placeholder for users of IEEE 802.11 that do not wish to implement the wired equivalent privacy (WEP) algorithms necessary for stronger authentication. Open system authentication allows the authentication frame exchange protocol to complete with a guaranteed result of "success." In this case, STA A would assert its identity to STA B, and STA B would respond with a successful result for the authentication. There is no verification of the identity of either STA. If any control over the STAs allowed to participate in the IEEE 802.11 WLAN is desired, this authentication algorithm should not be used.

The second authentication algorithm is the shared key authentication algorithm. This algorithm depends on both STAs having a copy of a shared WEP key. This algorithm uses the WEP encryption option to encrypt and decrypt a "challenge text" as the proof that the STAs share the same key. Beginning the authentication process, STA A sends its identity assertion to STA B. STA B responds to the assertion with an assertion of its own and a request to STA A to prove its identity by correctly encrypting the challenge text. STA A encrypts the challenge text (actually the entire frame body of the Authentication [management] frame) using the normal WEP encryption rules (including use of default and key mapping keys) and sends the result back to STA B. STA B decrypts the frame using the appropriate key and returns an Authentication frame to STA A with the success or failure of the authentication indicated. If the authentication is successful, the IEEE 802.11 standard says that each STA is authenticated to the other.

A STA may authenticate with any number of other STAs. The IEEE 802.11 standard does not place a limit on the number of authentications that may be processed. Therefore, a STA may preauthenticate with other STAs, even though there may be no immediate need for this connection. In examples later in this chapter, the desirability of preauthentication will be shown.

It should be noted that this algorithm really only authenticates STA A to STA B. The IEEE 802.11 Working Group believed that the AP somehow occupied a more privileged position than the mobile STAs when it came to authentication because it is always the mobile STA that initiates the authentication process. For this reason, it is only the mobile STA that performs the encryption operation on the challenge text. This limitation leaves the IEEE 802.11 WLAN open to some not-so-subtle security problems. In particular, a rogue AP could adopt the SSID of the ESS and announce its presence through the normal beaconing process. This action would cause mobile STAs to attempt to authenticate with the rogue. The rogue could always complete the authentication process with an indication of successful authentication. This response would cause mobile STAs to attempt to use the rogue for access to the WLAN. The rogue could then simply complete normal frame handshake procedures, and the mobile STAs would be the victims of a denial-of-service attack. A more active rogue could use more subtle means to attempt to gain access to the content of higher layer protocol frames containing user names, passwords, and other sensitive data. However, if the data are encrypted using WEP, it is highly unlikely that the rogue could successfully decrypt the information.

In 2004, IEEE 802.11i was approved. This amendment was developed specifically to address the shortcomings of the authentication methods defined in the IEEE 802.11 base standard. See Chapter 4 for additional information on IEEE 802.11i.

Association

Association is the mechanism through which IEEE 802.11 provides transparent mobility to STAs. Because IEEE 802.11 is designed to operate without wires tethering a piece of LAN equipment to a single location, support for mobility is built into the standard. Association is the process of a mobile STA "connecting" to an AP and requesting service from the WLAN. In the IEEE 802.11 base standard, association may be accomplished only after a successful authentication has been completed. Because open system authentication provides no significant benefit and shared key authentication is actually a way to leak information about the key, IEEE 802.11i made authentication before association optional.

When a mobile STA requests to be connected to the WLAN, it sends an Association Request frame to an AP. The association request includes information on the capabilities of the STA, such as the data rates it supports, the high-rate PHY options it supports, its contention-free capabilities, its support of WEP, and any request for contention-free services. The association request also includes information about the length of time that the STA may be in a low-power operating mode.

The information in an association request is used by the AP to decide whether to grant the association for the mobile STA. The policies and algorithms used by the AP to make this decision are not described in the standard. Some considerations are support for all of the required data rates and PHY options, requirements for contention-free services beyond the ability of the AP to support, long periods in low-power operation that require excessive buffer commitments from the AP, and the number of STAs currently associated. Because the IEEE 802.11 standard does not specify what information may be considered by the AP when deciding to grant an association, information not local to the AP may also be used, such as load balancing factors and availability of other APs nearby. When the AP responds to the mobile STA with an Association Response frame, the response includes a status indication. The status indication provides the mobile STA with the success or failure of the association request. If the request fails, the reason for that failure is in the status indication.

Once a STA is associated, the AP is responsible for forwarding data frames from the mobile STA toward their destination. It should be noted that the forwarding behavior of an IEEE 802.11 AP is not the same as for an IEEE 802.1 bridge. In IEEE 802.1 bridges, a frame received on one interface for which the bridge does not have an entry in its forwarding table is typically sent to all other interfaces on the bridge ("flooded"). In an 802.11 AP, the AP will forward frames to the DS only from an associated mobile STA. An AP will forward frames from the DS only when those frames are destined for an associated mobile STA. The destination of the data frames may be in the same BSS as the mobile STA, in which case the AP will simply transmit the data frame to the BSS, or it may be outside of the BSS. If the destination of a data frame is outside the BSS, the AP will send the frame into the DS. The use of the DS is outside of the scope of the IEEE 802.11 standard. However, most vendors of IEEE 802.11 equipment either bridge or route data frames from the BSS to the wired network serving as the DS. If the destination of the data frame is another mobile STA in a different BSS, the AP of the other BSS detects the frame on the DS and forwards it to the mobile STA. If the destination of the frame is entirely outside the ESS, the AP will forward the frame to the *portal*, i.e., the exit from the DS to the rest of the network. A portal is simply a transfer point between the wired LAN and the ESS, where frames *logically* enter the ESS. A portal may be an AP, a bridge, or a router. Because IEEE 802.11 is one of the family of IEEE 802 standards, an IEEE 802.11 frame must be translated from the IEEE 802.11 format to the format of the other LAN. This translation should be done according to IEEE Std 802.11 to another LAN. The entire IEEE 802.11 frame, including MAC header and FCS, should not be encapsulated within another MAC protocol.

Similarly, when a data frame is sent from outside the ESS to a mobile STA, the portal must forward the frame to the correct AP, i.e., the one that has the mobile STA associated in its BSS. The need for the portal to know where the mobile STAs are implies that there must be information kept in the DS allowing each mobile STA that is associated to be found at any given instant in time. Again, this function is not specified in the IEEE 802.11 standard, but is required of the DS to support the operation of the mobile STAs as they move from one AP to another.

Once a STA is successfully associated, it may begin exchanging data frames with the AP. Because the STA is mobile and also because the medium is subject to both slow and fast variations, the mobile STA will eventually lose contact with the AP. When this loss occurs, the mobile STA must begin a new association in order to continue exchanging data frames. Because the DS must maintain information about the location of each mobile STA and because data frames may have been sent to an AP with which the mobile STA no longer can communicate, a mobile STA will use a Reassociation Request frame after its initial association. The Reassociation Request frame includes all of the information in the original Association Request frame, plus the address of the last AP with which the STA was associated. The last AP's address allows the AP receiving the Reassociation Request frame to retrieve any frames at the old AP and deliver them to the mobile STA. Once the AP grants the reassociation, the mobile STA's older association is terminated. While this termination is outside the scope of the current IEEE 802.11 standard, the AP that has just granted the reassociation normally communicates with the AP with which the STA was last associated to cause the termination of the old association. The reason for this step should be clear. The association provides information to the DS about the location of the mobile STA. The mobile STA is allowed to have only one location in the ESS so that there is no ambiguity as to where frames destined for that mobile STA should be sent. This requirement is particularly true of a DS built from IEEE 802.1 bridges and switches, which cannot tolerate any ambiguity as to the direction that frames for a particular MAC address should be forwarded. Thus, the STA is permitted only a single association.

Address filtering (MAC function)

The address filtering mechanism in the IEEE 802.11 WLAN is a bit more complex than that of other IEEE 802 LANs. In a WLAN, it is not sufficient to make receive decisions on the DA alone. There may be more than one IEEE 802.11 WLAN operating in the same location and on the same medium and channel. In this case, the receiver must examine more than the DA to make correct receive decisions. IEEE 802.11 incorporates at least three addresses in every data and management frame that may be received by a STA. In addition to the DA, these frames also include the BSSID. A STA must use both the DA and the BSSID when making receive decisions, according to the IEEE 802.11 standard. This requirement ensures that the STA will discard frames sent from a BSS other than that with which it is associated. Filtering on the BSSID is particularly important to minimize the number of multicast frames with which the STA must deal.

Recent developments in IEEE 802.11 devices have introduced the *virtual* AP. A virtual AP is functionally the same as any other AP. The difference is that the hardware that supported a single BSS can now support a number of BSSs, each logically distinct from each other. Each of these BSSs requires a unique BSSID to allow STAs to see each virtual AP as identical to a physical AP.

Privacy (MAC function)

The privacy function is provided by the WEP mechanism described in Chapter 3. WEP allows the frames to be broadcast and received by any receiver because the content of the frames may be protected by a well-known encryption algorithm. Further details of the operation of WEP are provided in Chapter 3.

Power management

The power management mechanism is the most complex part of the IEEE 802.11 standard. It allows mobile STAs, in either an IBSS or an infrastructure BSS, to enter low-power modes of operation where they turn off their receiver and transmitter to conserve power. Because of the significant differences between the IBSS and infrastructure BSS, there are two different mechanisms for power management. One of the mechanisms is used only in IBSSs and the other only in infrastructure BSSs.

Power management in an IBSS

In an IBSS, power management is a fully distributed process, managed by the individual mobile STAs. Power management comprises two parts: the functions of the STA entering a low-power operating mode and the functions of the STAs that desire to communicate with that STA. For a STA to enter a low-power operating state, i.e., a state where it has turned off the receiver and transmitter to conserve power, the STA must successfully complete a data frame handshake with another STA with the power management bit set in the frame header. Any other STA is acceptable for this handshake. No requirement exists for the STA to complete this handshake with more than a single other STA, even though there may be many other STAs with which this STA has communicated in the BSS. If the STA has no data to send, it may use the Null Function [data] frame for this handshake. Until this frame handshake is completed, the STA must remain in the awake state. The IEEE 802.11 standard does not specify when a STA may enter or leave a low-power operating state, only how the transition is to take place.

Once the STA has successfully completed the frame handshake with the power management bit set, it may enter the power-saving state. In the power-saving state, the STA must wake up to receive every Beacon frame transmission. The STA must also stay awake for a period of time after each Beacon frame receipt, called the announcement (or ad hoc) traffic indication message (ATIM) window. The earliest the STA may reenter the power-saving state is at the conclusion of the ATIM window. The reason that a STA must remain awake during the ATIM window is that other STAs that are attempting to send frames to it will announce those frames during the ATIM window. If the power-saving STA receives an ATIM [management] frame, it must acknowledge that frame and remain awake until the end of the next ATIM window, following the next Beacon frame, in order to allow the other STA to send its data frame.

For a STA desiring to send a frame to another STA in an IBSS, the IEEE 802.11 standard requires that the sending STA estimate the power-saving state of the intended destination. The estimate of the power-saving state of another STA may be based on the last data frame received from the STA and on other information local to the sending STA. How the sending STA creates its estimate of the power-saving state of the intended destination is not described in the standard. If the sending STA determines that the intended destination is in the power-saving state, the sending STA may not transmit its frame until after it has received an acknowledgment of an ATIM frame, sent during the

ATIM window, from the intended destination. Once an acknowledgment of the ATIM is received, the STA will send the corresponding data frame after the conclusion of the ATIM window.

Multicast frames must also be announced by the sending STA during the ATIM window before they may be transmitted. The ATIM is sent to the same multicast address as the data frame that will be sent subsequently. Because the ATIM is sent to a multicast address, no acknowledgment will be generated, nor is one expected. Any STAs that wish to receive the announced multicast data frame must stay awake until the end of the next ATIM window, after the next Beacon frame. The STA sending the multicast data frame may send it at any time after the conclusion of the ATIM window.

This power management mechanism puts a slightly greater burden on the sending STA than on the receiving STA. Sending STAs must send an ATIM frame in addition to the data frame it desires to deliver to the destination. Sending STAs must buffer the frames to be sent to the power-saving destination until the destination awakens and acknowledges the ATIM frame. Because of the nature of the wireless medium, it may require several attempts before an ATIM frame is acknowledged. Each transmission of an ATIM frame consumes power at the sending STA. The receiving STA must awaken for every Beacon frame and ATIM window, but need not make any transmissions unless it receives an ATIM frame. This power management mechanism allows reasonable power savings in all mobile STAs. However, there is a minimum duty cycle required of both senders and destinations, in the ratio of the time of the ATIM window to the time of the beacon period, which limits the maximum power savings that may be achieved. During the development of the IEEE 802.11 standard, the limitation of the maximum power savings was thought to be a reasonable trade-off for the complexity involved.

Power management in an infrastructure BSS

In an infrastructure BSS, the power management mechanism is centralized in the AP. This power management mechanism allows much greater power savings for mobile STAs than does the mechanism used in IBSSs. The reason is that the AP assumes all of the burden of buffering data frames for power-saving STAs and delivering them when the STAs requests. This approach allows the mobile STAs to remain in their power-saving state for much longer periods.

The responsibilities of the mobile STAs in an infrastructure BSS are to inform the AP, in its Association Request frame, of the number of beacon periods that the STA will be in its power-saving mode, to awaken at the expected time of a Beacon frame transmission to learn whether any data frames are waiting, and to complete a successful frame handshake with the AP, while the power management bit is set, to inform the AP when the STA will enter the powersaving mode. A mobile STA can achieve much deeper power savings than in the IBSS because it is not required to awaken for every Beacon frame nor to stay awake for any length of time after the Beacon frames for which it does awaken. The mobile STA must also awaken at times determined by the AP when multicast frames are to be delivered. This time is indicated in the Beacon frames as the delivery traffic indication map (DTIM). If the mobile STA is to receive multicast frames, it must be awake at every DTIM.

The AP will buffer data frames for each power-saving STA with which it has associated. It will also buffer all multicast frames if it any associated STAs that are in the power-saving mode. The data frames will remain buffered at the AP for a minimum time not less than the number of beacon periods indicated in the mobile STA's Association Request frame. The IEEE 802.11 standard indicates that an AP may use an aging algorithm to discard buffered frames that are older than it is required to preserve, although a specific algorithm is not described. An AP might also discard frames if it has run out of physical buffer capacity to hold them. Once the AP has frames buffered for a power-saving STA, it will indicate this fact in the traffic indication map (TIM) sent with each Beacon frame. Every STA that is associated with the AP is assigned an association ID (AID) during the association process. The AID indicates a single bit in the TIM that reflects the status of frames buffered for that STA. When the bit in the TIM is set, at least one frame is buffered for the corresponding STA. When the bit is clear, no frames are buffered for the corresponding STA. A special AID, AID zero, is dedicated to indicating the status of buffered multicast traffic. The AP will send the TIM, updated with the latest buffer status, with every Beacon frame.

If an AP has any buffered multicast frames, those frames are sent immediately after the Beacon frame announcing the DTIM. If more than one multicast frame is to be sent, the AP will indicate this fact by setting the More Data bit in the Frame Control field of each multicast frame except for the last to be sent. Following the transmission of any buffered multicast frames, the AP will send frames to active STAs and to STAs that have requested the delivery of frames buffered for them. A mobile STA requests delivery of buffered frames by sending a PS-Poll frame to the AP. The AP will respond to each PS-Poll frame by sending one frame to the requesting STA. In the Frame Control field of the frame sent in response to the PS-Poll frame, the AP will set the More Data bit if there are more frames buffered for the STA. The STA is required to send a PS-Poll frame to the AP for each data frame it receives with the More Data bit set. This step ensures that the STA will empty the AP's buffer of the frames the AP is holding for the STA. The IEEE 802.11 standard does not state any time requirement for the STA to send the PS-Poll frame after seeing the More Data bit. Thus, some implementations may rapidly retrieve all buffered frames from the AP, and others may operate at a much more leisurely pace.

An AP that is also a point coordinator (PC) running a contention-free period (CFP) will use the CFP to deliver buffered frames to STAs that are CF pollable. It may also use the CFP to deliver multicast frames after the DTIM is announced.

Synchronization

Synchronization is the process in which the STAs in a BSS get in step with each other so that reliable communication is possible. The MAC provides the synchronization mechanism to allow support of PHYs that make use of frequency hopping or other time-based mechanisms where the parameters of the PHY change with time. The process involves beaconing, to announce the presence of a BSS, and scanning, to find a BSS. Once a BSS is found, a STA joins the BSS. This process is entirely distributed, in both IBSSs and infrastructure BSSs, and relies on a common time base provided by a timer synchronization function (TSF).

The TSF maintains a 64-bit timer running at 1 MHz and updated by information from other STAs. The tolerance of the time is 25 ppm. When a STA begins operation, it resets the timer to zero. The timer may be updated by information received in Beacon frames, as described below.

Timer synchronization in an infrastructure BSS

In an infrastructure BSS, the AP is responsible for transmitting a Beacon frame periodically. The time between Beacon frames is called the *beacon period* and is included as part of the information in the Beacon frame in order to inform STAs receiving the Beacon frames when to expect the next Beacon frame. The AP will attempt to transmit the Beacon frame at the target beacon transmission time (TBTT), when the value of the TSF timer of the AP, modulo the beacon period, is zero. The Beacon frame, however, is a frame like any other and is sent using the same rules for accessing the medium. Thus, the Beacon frame may be delayed beyond the TBTT because other traffic occupies the medium and backoff delays occur. In addition, because the Beacon frame is sent to the broadcast address, it will not be retransmitted should the frame be corrupted. As a result, the Beacon frame may not be received by some, or all, of the STAs in the BSS. This consequence is expected as a result of the normal operation of the IEEE 802.11 WLAN and does not result in any degradation of the operation of the LAN. Without regard to whether the Beacon frame was sent at the TBTT or whether it was corrupted, the AP will attempt to transmit the following Beacon frame at the next TBTT. In a lightly loaded BSS, the Beacon frame will usually be sent at the TBTT and be spaced apart by exactly the beacon period. As the load in the BSS increases, the Beacon frame will be delayed beyond the TBTT more often.

The TSF timer in an AP is reset to zero upon initialization of the AP and is then incremented by the 1 MHz clock of the AP. At the time of each Beacon frame transmission, the current value of the timer is inserted in the Beacon frame. For a mobile STA in an infrastructure BSS, the synchronization function is very simple. A mobile STA will update its TSF timer with the value of the timer it receives from the AP in the Beacon frame, modified by any processing time required to perform the update operation. Thus, the timer values in all of the mobile STAs in the BSS receiving the Beacon frame are synchronized to timer value of the AP.

Timer synchronization in an IBSS

In an IBSS, there is no AP to act as the central time source for the BSS. In an IBSS, the timer synchronization mechanism is completely distributed among the mobile STAs of the BSS. Because there is no AP, the mobile STA that starts the BSS will begin by resetting its TSF timer to zero and transmitting a Beacon frame, choosing a beacon period. This step establishes the basic beaconing process for this BSS. After the BSS has been established, each STA in the IBSS will attempt to send a Beacon frame after the TBTT arrives. To be sure that at least one Beacon frame is sent in each beacon period and to minimize actual collisions of the transmitted Beacon frames on the medium, each STA in the BSS will choose a random delay value that it will allow to expire after the TBTT before it attempts its Beacon frame transmission. If the STA receives a Beacon frame from another STA in the BSS before the delay expires, the receiving STA's Beacon frame transmission will be canceled. If, however, the delay expires without the STA receiving a Beacon frame, the Beacon frame transmission will proceed. It is easy to see that more than one transmission may occur simultaneously and cause corruption of the transmission for some receivers and good reception for others. Thus, some receivers may receive more than one Beacon frame in a single beacon period. This operation is allowed in the IEEE 802.11 standard and does not cause any degradation or confusion in the receiving STAs.

Beaconing also interacts with power management in the IBSS. The IEEE 802.11 standard requires that the STA, or STAs, that send a Beacon frame must not enter the power save state until they receive a Beacon frame from another STA in the BSS. This restriction on the beaconing STA is to ensure that there is at least one STA in the IBSS awake and able to respond to Probe Request frames.

Because each STA will send its own value for the TSF timer in the Beacon frames it transmits, the rules for updating the TSF timer in a STA in an IBSS are slightly more complex than the rules for STAs in an infrastructure BSS. In an IBSS, a STA will update its TSF timer with the value of a received Beacon frame if the received value, after modifying it for processing times, is greater than the value currently in the timer. If the received value is not greater than the local timer value, the received value is discarded. The effect of this selective updating of the TSF timer and the distributed nature of beaconing in an IBSS is to spread the value of the TSF timer of the STA with the fastest running clock throughout the BSS. The reason for this algorithm is to prevent time from running backwards at any STA in the IBSS. The speed with which the fastest timer value spreads is dependent on the number of STAs in the BSS and whether all STAs are able to communicate directly. If the number of STAs is small and all STAs can communicate directly, the timers of all STAs will be updated with the fastest timer value with a period proportional to the number of STAs in the BSS. As the number of STAs grows and collision of Beacon frame transmissions occurs, the spread of the fastest timer value will slow. Similarly, if all STAs cannot communicate directly, it requires more than one STA to propagate the fastest timer value to the outlying reaches of the BSS. Thus, the spread of the fastest timer value slows proportional to the number of hops it must take to reach all STAs.

Synchronization with FH PHYs

The MAC provides the capability to support FH PHYs through its synchronization mechanism. Similar to beaconing, changes in a FH PHY (i.e., movements to other channels) occur periodically (i.e., the dwell period). With the TSF timers of all STAs roughly synchronized, all STAs in a BSS will make these changes simultaneously. This coordination minimizes the time that could be lost during resynchronization. In particular, all STAs in a BSS will change to the new channel when the TSF timer value, modulo the dwell period, is zero.

Scanning

In order for a mobile STA to communicate with other mobile STAs in an IBSS or with the AP in an infrastructure BSS, it must first find the STAs or APs. The process of finding another STA or AP is *scanning*. Scanning may be either passive or active. Passive scanning involves only listening for IEEE 802.11 traffic. Active scanning requires the scanning STA to transmit and elicit responses from IEEE 802.11 STAs and APs. Both methods are described in the IEEE 802.11 standard. The use of one method or the other is left to the implementer.

Passive scanning allows a mobile STA to find a BSS, while minimizing the power expended, by not transmitting, only listening. The process a STA uses for passive scanning is to move to a channel, listen for Beacon and Probe Response frames, and extract a description of a BSS from each of these frames received. After a period of time, the STA changes to a different channel and listens again. At the conclusion of the passive scan, which may involve listening to one or more channels, the STA has accumulated information about the BSSs that are in its vicinity. Although this scanning method may reduce the power expended by the STA while scanning, the cost is the additional time required to listen for frames that may not occur because there is no BSS on the current channel.

Active scanning allows a mobile STA to find a BSS, while minimizing the time spent scanning, by actively transmitting queries that elicit responses from STAs in a BSS. In an active scan, the mobile STA will move to a channel and transmit a Probe Request frame. If a BSS is on the channel that matches the SSID in the Probe Request frame, a STA in that BSS will respond by sending a Probe Response frame to the scanning STA. This responding STA is the AP in an infrastructure BSS and the last STA to send a Beacon frame in an IBSS. The Probe Response frame includes the information necessary for the scanning STA to extract a description of the BSS. The scanning STA will also process any gratuitously received Probe Response and Beacon frames. Once the scanning STA has processed any responses, or has decided there will be no responses, it may change to another channel and repeat the process. At the conclusion of the scan, the STA has accumulated information about the BSSs in its vicinity.

Although the IEEE 802.11 standard describes both active and passive scanning, it does not state any requirements on when each method is to be used. Thus, vendors of IEEE 802.11 equipment are free to innovate and create their own policies regarding the use of active and passive scanning. Some methods that may be seen are to begin a scan with active scanning to rapidly find any IEEE 802.11 WLAN in the vicinity and to revert to passive scanning to conserve power if the active scan is not successful.

Joining a BSS

Once a STA has performed a scan that results in one or more BSS descriptions, the STA may choose to join one of the BSSs. The joining process is a purely local process that occurs entirely internal to the IEEE 802.11 mobile STA. There is no indication to the outside world that a STA has joined a particular BSS. While the IEEE 802.11 standard does describe what is required of a STA to join a BSS, it does not describe how a STA should choose one BSS over another.

Joining a BSS requires that all of the mobile STA's MAC and PHY parameters be synchronized with the desired BSS. To accomplish this step, the STA must update its TSF timer with the value of the timer from the BSS description, modified by adding the time elapsed since the description was acquired. This step will synchronize the TSF timer to the BSS. It will also, coincidentally, synchronize the hopping of FH PHYs. The STA must also adopt the PHY parameters in the FH Parameter Set and/or the DS Parameter Set information elements, as well as the required data rates. This step will ensure that the PHY is operating on the same channel as the rest of the STAs in the BSS. The BSSID of the BSS must be adopted, as well as the parameters in the Capability Information field, such as WEP and the IEEE 802.11b high-rate PHY capabilities. The beacon period and DTIM period must also be adopted. Once this process is complete, the mobile STA has joined the BSS and is ready to begin communicating with the STAs in the BSS.

COMBINING MANAGEMENT TOOLS

The MAC management tools described in this chapter need not be used in isolation. They are most powerful when used in combination.

Combining power-saving periods with scanning

One of the most useful combinations of MAC management tools is combining power-saving periods and scanning in an infrastructure network. With this combination, a mobile STA would complete the frame handshake with its AP to inform the AP that the STA would be entering the power-saving mode. The AP would then begin buffering any arriving data frames for the mobile STA. Then, instead of entering the power-saving mode, the mobile STA would perform active or passive scanning for a period of time to gather BSS descriptions of other BSSs in the vicinity. The mobile STA would then rejoin the BSS where it is associated before either the DTIM approached or the number of beacon periods elapsed when the AP might begin discarding frames buffered for the mobile STA.

This combination allows a mobile STA to gather information about its environment, i.e., the other BSSs that are nearby, while it has the luxury of being associated and in communication with an AP. Being associated with an AP is a much better time for scanning than after communication with the AP is lost. Then, when the mobile STA eventually does move out of communication with its AP, it has all of the necessary information to quickly verify that communication with one of the other BSSs is possible and then to authenticate and reassociate with the new BSS. This sequence of events minimizes the disruption of communication when it is necessary for a mobile STA to roam from one BSS to another.

Preauthentication

Another useful combination of MAC management tools provides preauthentication. Here a mobile STA combines scanning with authentication. As the mobile STA scans for other BSSs, it will initiate an authentication when it finds a new BSS. This step also reduces the time required for a STA to resume communication with a new BSS, once it loses communication with the BSS with which it was associated. Now, when communication with the current BSS is no longer satisfactory, the mobile STA can join the new BSS and simply reassociate with the AP. The authentication was completed while the mobile STA was scanning, some time in the past. It should be noted that some vendors might choose to propagate a STA's authentication from one AP to another through the DS, obviating the need for more than a single, initial authentication. The IEEE 802.11 standard does not discuss this operation, nor does it prohibit it.

AREAS FOR IMPROVEMENT

Since the first edition of this handbook was published, there has been a lot of time to observe the behaviors of many 802.11 STA implementations. This section is devoted to some advice to improve new implementations.

Scanning and roaming

Many implementations have very poor scanning and roaming behavior. Most of them have room for significant improvement. One significant area of concern is the algorithms used for active scanning. For example, many current implementations scan at a constant rate, regardless of the quality of the link between the STA and AP. This practice is needlessly wasteful of bandwidth because there is little need to scan when the link to the AP is of high quality. The continual transmission of Probe Response frames from the APs consumes a small, but significant, amount of the channel's capacity.

One improvement, therefore, would be to use an adaptive algorithm to determine the rate of scanning. Adaptive rate scanning will maintain a very low rate of sending Probe Request frames when the quality of the link between a STA and AP is very good. As the link quality decreases, the rate of sending Probe Request frames increases, so that the STA will have a very good list of AP candidates to which it might roam when a threshold is reached that indicates the current AP is no longer acceptable.

Another area for improvement in STA active scanning algorithms is hysteresis. Many current implementations make a decision to roam to a new AP as soon as a candidate is found that looks better in some respect than the AP where the STA is currently associated. This movement causes *ping ponging* between two APs in many cases, with little or no benefit to the STA. In fact, the time spent associating and authenticating can actually reduce the STA's throughput and cause an excessive number of lost frames.

A third area for improvement in scanning is to use directed probe requests, rather than broadcast probe requests. When the SSID is not the wild card SSID, i.e., the probe request is directed, only the APs on the WLAN to which the STA is sending the Probe Request frames will respond. This approach can significantly reduce the bandwidth consumed by Probe Response frames in an area where more than one WLAN is available. An even further improvement in this area is to send the Probe Request frame to a specific, individual BSSID. This approach reduces the APs that will respond to the Probe Request frame to only a single AP and can be used when checking the current quality of the link to APs that have been found by earlier scans.

Use of status and reason codes

If there is one egregious problem with current implementations, it is the lack of use of the status codes and reason codes provided in the IEEE 802.11 standard. Most APs send only two values for status codes: 0 for success and 1 for all failures. Most STAs check only to see whether the status code is nonzero. If the status code is nonzero, most STAs execute exactly the same algorithm, on exactly the same data, to come to exactly the same conclusion that resulted in the delivery of the nonzero status code in the first place. Most often, this process results in a STA that appears to be *locked up*, i.e., not providing any connectivity to the WLAN.

The solution to this problem, of course, is for APs to use the status codes defined in the IEEE 802.11 standard and send them to the STA when the appropriate conditions apply. "Unspecified error" (status code 1) should be used only if there is no other code that better describes the reason for a failed association or authentication. Using the proper status code values (and reason code values in Disassociation and Deauthentication frames) will provide information to the STA to allow it to, perhaps, correct the error. At least the status code values can be observed in a frame capture (sniffer) or log file to provide some insight into the cause of an apparent protocol failure between STA and AP.

The AP sending useful status code values is of little benefit if the STA does not recognize them and take action, based on the individual status or reason code value returned, to deal with the failure that is indicated. A STA must use the information that the AP provides it. Using this information can significantly improve the behavior of the STA and provide a much better user experience. The status and reason code values are described in Chapter 3.

Chapter 10 MAC management information base (MIB)

The IEEE 802.11 MIB is an object managed by Simple Network Management Protocol version 2 (SNMPv2). The MIB contains a number of configuration controls, option selectors, counters, and status indicators that allow an external management agent to determine the status and configuration of an IEEE 802.11 STA as well as to probe the agent's performance and tune its operation. The MIB in a STA comprises two major sections: one for the MAC and one for the PHY. The PHY section is subdivided into subsections that are specific to each PHY. In the MIB definition, there is also a compliance section that describes the required portions of the MIB and the parts that are optional. All of the attributes are arranged in tables, coordinating the attributes that are related to a single function.

The MAC MIB comprises two sections: the STA management attributes and the MAC attributes. The STA management attributes are associated with the configuration of options in the MAC and the operation of MAC management. The MAC attributes are associated with the operation of the MAC and its performance.

STA MANAGEMENT ATTRIBUTES

The STA management attributes configure and control the operation of the options of the IEEE 802.11 MAC as well as assist in the management of the STA.

The dot11StationID attribute is 48 bits long and is designed to allow an external manager to assign its own identifier to a STA for the sole purpose of managing the STA. This attribute does not change the actual MAC address of the STA. Its default value is the unique MAC address of the STA.

The dot11MediumOccupancyLimit attribute provides a limit to the amount of time that the PC may control access to the medium. After this limit is reached, the PC must relinquish control of the medium to the DCF, allowing at least enough time to transmit a single maximum-length MPDU, with fragmentation, before taking control of the medium again. The units used by this attribute are the TU, or 1024 μs. The default value of this attribute is 100 TU. The maximum value is 1000 TU, or 1.024 s. This attribute may be changed by an external manager to allocate more, or less, bandwidth to the PC.

The dot11CFPollable attribute is a Boolean flag that indicates the capability of the STA to respond to the CF-Poll frame. This attribute is read-only. It may not be changed by an external manager.

The dot11CFPPeriod attribute defines the length of the CFP, in units of the DTIM interval, which, in turn, is in units of the beacon period, which is measured in TU. Thus, to determine the time value of the CFP, multiply the value of this attribute by the DTIM interval (found in the Beacon frame) and multiply the result by 1024 μs. This calculation shows the relationship between Beacon frames, DTIMs, and the CFP.

The dot11CFPMaxDuration attribute would appear to have an identical function to the dot11MediumOccupancyLimit attribute. However, this attribute is modified by the MLME-Start.request primitive that is used to initiate a BSS. There seems to be no reason for the difference in the maximum allowable values for this attribute and dot11MediumOccupancyLimit.

The dot11AuthenticationResponseTimeout attribute places an upper limit, in TU, on the time that a STA is allowed to wait before the next frame in an authentication sequence is determined not to be forthcoming. When this timeout expires, the authentication is judged to have failed.

The dot11PrivacyOptionImplemented attribute is a Boolean indicator of the presence of the privacy option. This attribute simply indicates that the option is implemented. It does not indicate whether WEP is in use.

The dot11PowerManagementMode attribute indicates the state of power management in the STA. This attribute is very likely to indicate to an external manager that the STA is always in the active mode, never in the power-saving mode. The reason is that the STA must be in the active mode to receive a frame from the external manager. Unless a significant time delay occurs between the reception of the frame from the external manager and the determination of the value of this attribute, the STA is likely still to be in the active mode. While this attribute may not be useful to an external manager, a local manager can determine the ratio of active-to-power-saving time in a STA through polling this attribute.

The dot11DesiredSSID attribute indicates the SSID used during the latest scan operation by the STA. Normally, the value of this attribute will be the same as the SSID of the IEEE 802.11 WLAN with which the STA is associated. An external manager may change this value should it be desirable for the STA to begin looking for a different WLAN.

The dot11DesiredBSSType attribute indicates the type of BSS that the STA sought during the latest scan operation. This attribute may be set by an external manager to force scanning for a particular BSS type.

The dot11OperationalRateSet attribute is a list of data rates that may be used by the STA to transmit in the BSS with which it is associated. The STA must also support reception at the rates indicated in this attribute. The rates in this attribute are a subset of the rates in the dot11SupportedRates in the PHY section of the MIB.

The dot11BeaconPeriod attribute controls the time that elapses between target Beacon frame transmission times. This attribute is set by the MLME-Start.request primitive. It may be changed by an external manager. However, any change to this attribute will require that any current BSS be dissolved and a new BSS started with the new beacon period. There is no provision in the IEEE 802.11 standard to gracefully change the beacon period once the BSS has been established.

The dot11DTIMPeriod attribute controls the number of beacon periods that elapse between DTIMs. This attribute is also set by the MLME-Start.request primitive. While it may be changed by an external manager, any change will be ineffective until the current BSS is dissolved and a new BSS is started.

The dot11AssociationResponseTimeout attribute places an upper limit on the amount of time that a STA will wait for a response to its association request. If an association response is not received before the timeout expires, the request is judged to have failed.

The dot11DisassociateReason attribute indicates the reason code received in the most recently received Disassociation frame. In combination with dot11DisassociateStation, an external manager can track the location and reasons that STAs are disassociated from the WLAN. The dot11DeauthenticateReason and dot11DeauthenticateStation attributes are used similarly to track deauthentications in the WLAN. The dot11AuthenticateFailReason and dot11AuthenticateFailStation attributes provide similar information about failures during the authentication process. If a large number of STAs are indicating authentication failures, deauthentications, or disassociations, this situation may be an indication that an AP is misbehaving or that an attack is in progress against the WLAN.

The dot11AuthenticationAlgorithm attribute is an entry in a table that holds an entry for each authentication algorithm supported by the STA. Every STA must support the open system algorithm. If the STA also supports the shared key algorithm, the table will hold an entry for that algorithm. Corresponding to each algorithm entry in the table is the dot11AuthenticationAlgorithms-Enable attribute. This attribute indicates whether the associated authentication algorithm is enabled, i.e., if it may be used for authentication by this STA.

The dot11WEPDefaultKeyValue attribute holds one of the WEP default keys. There may be as many as four default keys in a table in the STA. This attribute is intended to be write-only. The IEEE 802.11 standard specifies that reading this attribute shall return a value of zero or null.

There is a table of attributes for the WEP key mapping keys. This table holds three accessible attributes: dot11KeyMappingAddress, dot11KeyMapping-WEPOn, and dot11KeyMappingValue.

The dot11KeyMappingAddress attribute holds the MAC address of a STA with which there exists a key mapping relationship. The dot11KeyMapping-WEPOn attribute is a Boolean value and indicates whether the key mapping key is to be used when communicating with the STA with the corresponding address. The dot11KeyMappingValue attribute is the key to be used when key mapping must be used to communicate with the STA with the corresponding address. There is one entry in the key mapping table, consisting of these three attributes, for each STA for which a key mapping relationship exists.

The dot11PrivacyInvoked attribute is a Boolean attribute that indicates when WEP is to be used to protect data frames. When true, all data frames are to be encrypted using either a default key or a key mapping key before they are transmitted. This value may be changed by an external manager.

The dot11WEPDefaultKeyID attribute identifies which of the four default keys are to be used when encrypting data frames with a default key. Choosing to use a default key for which a corresponding value has not been set will result in an error being returned to higher layer protocols to indicate that an attempt was made to encrypt with a null key.

The dot11WEPKeyMappingLength attribute indicates the number of entries that may be held in the key mapping table. The minimum value for this attribute is 10, indicating that the key mapping table must hold at least 10 entries.

The dot11ExcludeUnencrypted attribute is a Boolean attribute that controls whether a STA will receive unencrypted data frames. When this attribute is true, only received data frames that were encrypted will be indicated to higher layer protocols. Unencrypted data frames will be discarded. When an unencrypted data frame is discarded, the value of dot11WEPExcludedCount is incremented. If the dot11WEPExcludedCount is increasing rapidly, it may be due to a STA that is misconfigured and attempting to exchange frames without encryption.

The dot11WEPICVErrorCount attribute tracks the number of encrypted frames that have been received and decrypted, but for which the ICV indicates the decryption was not successful. This counter can indicate when an excessive number of decryption errors are encountered. Such a situation may be due to a failure to use the same key as the key with which the message was encrypted, possibly due to a key update that was missed.

The STA management portion of the MIB also includes three notification objects, corresponding to three occurrences that are usually exceptional. The dot11Disassociate object is activated when a STA receives a Disassociation frame. The dot11Deauthenticate object is activated when the STA receives a Deauthentication frame. The dot11AuthenticateFail object is activated when the STA does not complete an authentication sequence successfully. Each of these notifications can be useful to both local and remote management agents.

MAC ATTRIBUTES

The MAC attributes tune the performance of the MAC protocol, monitor the performance of the MAC, identify the multicast addresses that the MAC will receive, and provide identification of the MAC implementation.

The dot11MACAddress attribute is the unique, individual address of the MAC. It is this address that the MAC considers to be its own and for which it will pass received frames to higher layer protocols. The default for this address is the manufacturer-assigned, globally administered 48-bit MAC address. This attribute may be changed by a local or external manager.

The dot11RTSThreshold attribute controls the transmission of RTS [control] frames prior to data and management frames. The value of this attribute defines the length of the smallest frame for which the transmission of RTS is required. Frames of a length less than the value of this attribute will not be preceded by RTS. The default value of this attribute is 2347, effectively turning off the transmission of RTS for all frames. This attribute may be changed by a local or external manager should it be desirable to enable the transmission of RTS. The IEEE 802.11 standard does not provide any guidelines for when to modify this attribute. However, if the MAC counters indicating frame errors and retransmissions are increasing rapidly, enabling RTS may address this situation.

The dot11ShortRetryLimit attribute controls the number of times a frame that is shorter than the dot11RTSThreshold attribute value will be transmitted without receiving an acknowledgment before that frame is abandoned and a failure is indicated to higher layer protocols. The default value of this attribute is 7. It may be modified by local and external managers.

The dot11LongRetryLimit attribute controls the number of times a frame that is equal to or longer than the dot11RTSThreshold attribute value will be transmitted without receiving an acknowledgment before that frame is abandoned and a failure is indicated to higher layer protocols. The default value of this attribute is 4. It may be modified by local and external managers.

The dot11FragmentationThreshold attribute defines the length of the largest frame that the PHY will accept. Frames larger than this threshold must be fragmented. The default value of this attribute is dependent on the PHY parameter aMPDUMaxLength. If the value of aMPDUMaxLength is greater than or equal to 2346, the default value is 2346. If the value of aMPDUMax-Length is less than 2346, the default value is aMPDUMaxLength. The value of this attribute tracks the value of aMPDUMaxLength. If aMPDUMax-Length becomes less than the value of this attribute during operation of the STA, this attribute will be set to the value of aMPDUMaxLength. This attribute may be modified by local and external managers, but may never exceed the value of aMPDUMaxLength.

The dot11MaxTransmitMSDULifetime attribute controls the length of time that MSDU transmission attempts will continue after the initial transmission attempt. Because a frame may be fragmented and the retry limits apply to only a single frame of the fragment stream, this timer limits the amount of bandwidth that may be consumed while attempting to deliver a single MSDU. The value of this attribute is in TUs. The default value is 512, or approximately 524 ms. This attribute may be modified by local and external managers.

The dot11MaxReceiveLifetime attribute controls the length of time that a partial fragment stream will be held pending reception of the remaining fragments necessary for complete reassembly of the MSDU. If the entire set of fragments has not been received before the lifetime expires, the fragments already received will be discarded. The default value of this attribute, measured in TUs, is 512. The value of this attribute may be modified by local and external managers.

The dot11ManufacturerID attribute is a variable-length character string that identifies the manufacturer of the MAC. This attribute may contain other information, at the manufacturer's discretion, up to the maximum of 128 characters.

The dot11ProductID attribute is a variable-length character string that identifies the MAC. This attribute may contain other information, at the manufacturer's discretion, up to the maximum of 128 characters.

The dot11TransmittedFragmentCount attribute is a counter that tracks the number of successfully transmitted fragments. As far as this counter is concerned, an MSDU that fits in a single frame without fragmentation is also considered a fragment and will increment this counter. A successful transmission is an acknowledged data frame sent to an individual address or any data or management frame sent to a multicast address.

The dot11MulticastTransmittedFrameCount attribute is a counter that tracks only transmitted multicast frames. This counter is incremented for every frame transmitted with the group bit set in the destination MAC address.

The dot11FailedCount attribute is a counter that tracks the number of frame transmissions that are abandoned because they have exceeded either the dot11ShortRetryLimit or dot11LongRetyLimit value. This counter, along with the retry and multiple retry counters, can provide an indication of the condition of a BSS. As the load in a BSS increases or the error rate of the medium increases, these counters will increment more rapidly.

The dot11RetryCount attribute is a counter that tracks the number of frames that required at least one retransmission before being delivered successfully.

The dot11MultipleRetryCount attribute is a counter that tracks the number of frames that required more than one retransmission to be delivered successfully.

These first five counters can provide additional counter information. The number of individually addressed frames transmitted is equal to the dot11MulticastTransmittedFrameCount value subtracted from the dot11TransmittedFragementCount value. The number of frames delivered successfully after only one retransmission is equal to the dot11MultipleRetryCount value subtracted from the dot11RetryCount value. The number of frames delivered successfully on the first transmission attempt is equal to the dot11RetryCount value subtracted from the number of individually addressed frames transmitted.

The dot11FrameDuplicateCount attribute is a counter that tracks the number of duplicate frames received. The value of this counter is indicative of the number of acknowledgments that failed to be delivered.

The dot11RTSSuccessCount attribute is a counter that increments for each Clear to Send (CTS) frame received in response to a Request to Send (RTS) frame.

The dot11RTSFailureCount attribute is a counter that increments each time a CTS is not received in response to an RTS.

The dot11ACKFailureCount attribute is a counter that tracks the number of times a data or management frame is sent to an individual address and does not result in the reception of an ACK frame from the destination.

The dot11ReceivedFragmentCount attribute is a counter that tracks the number of fragments received. A fragment is any received data or management frame for the purpose of this counter.

The dot11MulticastReceivedCount attribute is a counter that tracks the number of frames received by the STA that match a multicast address in the group addresses table or were sent to the broadcast address.

The dot11FCSErrorCount attribute is a counter that tracks the number of frames received, of any type, that resulted in an FCS error. This counter provides another indication of the condition of the BSS. Increasing load and increasing error rate will both result in this counter increasing more rapidly.

The dot11TransmittedFrameCount attribute is a counter that tracks the number of MSDUs that have been transmitted successfully. This counter increments only if the entire fragment stream required to transmit an MSDU is sent and an acknowledgment is received for every fragment.

The dot11WEPUndecryptableCount attribute is a counter that tracks the number of frames received without FCS errors and with the Protected Frame bit (formerly WEP bit) indicating that the frame is encrypted, but that cannot be decrypted due to the dot11WEPOn indicating a key mapping key is not valid or the STA not implementing WEP. When this counter increments, it indicates that the receiving STA is misconfigured, has somehow gotten into a BSS that requires WEP, or has missed a key update for a key mapping STA.

Multicast addresses for which the STA will receive the frame and indicate it to higher layer protocols are stored in an instance of the dot11Address attribute.

This attribute is one entry in the dot11GroupAddressesTable. The table is dynamic, and entries may be added to and deleted from the table at any time.

The dot11ResourceTypeIDName attribute is required by IEEE 802.1F. It is a read-only, fixed-length character string. Its default value is RTID.

The dot11ResounceInfoTable contains four more attributes required by IEEE 802.1F. These attributes are dot11manufacturerOUI, dot11manufacturer-Name, dot11manufacturerProductName, and dot11manufacturerProduct-Version. All of these attributes are read-only. The dot11manufacturerOUI attribute contains the IEEE-assigned 24-bit organizational unique identifier that forms half of a globally administered MAC address. The dot11-manufacturerName attribute is a variable-length character string containing the name of the manufacturer of the MAC. The dot11manufacturerProduct-Name attribute is also a variable-length character string containing the product-identifying information for the MAC. The dot11manufacturer-ProductVersion attribute is also a variable-length character string that identifies the version information for the MAC.

Chapter 11 **The physical layer (PHY)**

PHY FUNCTIONALITY

At the bottom of the Open System Interconnection (OSI) stack is the PHY (see Figure 11–1). The PHY is the interface between the MAC and wireless media. It transmits and receives data frames over a shared wireless media. The PHY provides three levels of functionality. First, the PHY provides a frame exchange between the MAC and PHY under the control of the physical layer convergence procedure (PLCP) sublayer. Second, the PHY uses signal carrier and spread spectrum modulation to transmit data frames over the media under the control of the physical medium dependent (PMD) sublayer. Third, the PHY provides a carrier sense indication back to the MAC to verify activity on the media.

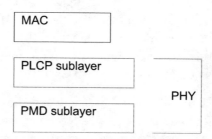

Figure 11–1: The OSI model

Each of the PHYs is unique in terms of modulation type and designed to coexist with each other and operate with the MAC described in Chapter 3. While developing the IEEE 802.11 standard, the specifications for IEEE 802.11 were selected to meet the RF emissions guidelines specified by the Federal Communications Commission (FCC), European Telecommunications Standards Institute (ETSI), and Ministry of Public Management, Home Affairs, Posts and Telecommunications (MPHPT) for Japan. This chapter provides an overview and gives the reader a basic understanding of the key specifications for each PHY.

DIRECT SEQUENCE SPREAD SPECTRUM (DSSS) PHY

The DSSS PHY is one of the three initial PHYs supported in the IEEE 802.11 standard and uses the 2.4 GHz frequency band as the RF transmission media. Data transmission over the media is controlled by the DSSS PMD sublayer as directed by the DSSS PLCP sublayer. The DSSS PMD takes the binary bits of information from the PLCP protocol data unit (PPDU) and transforms them into RF signals for the wireless media by using carrier modulation and DSSS techniques. Figure 11–2 illustrates the basic of elements of the DSSS PMD transmitter and receiver.

DSSS PLCP sublayer

The PPDU is unique to the DSSS PHY. The PPDU frame consists of a PLCP preamble, PLCP header, and MAC protocol data unit (MPDU) (see Figure 11–3). The receiver uses the PLCP preamble to acquire the incoming signal and synchronize the demodulator. The PLCP header contains information about MPDU from the sending DSSS PHY. The PLCP preamble and PLCP header are always transmitted at 1 Mbit/s, and the MPDU can be sent at either 1 Mbit/s or 2 Mbit/s.

The SYNC field is 128 bits (symbols) in length and contains a string of ones that are scrambled prior to transmission. The receiver uses this field to acquire the incoming signal and synchronize the receiver's carrier tracking and timing prior to receiving the start of frame delimiter (SFD).

The SFD field contains information marking the start of a PPDU frame. The SFD specified is common for all IEEE 802.11 DSSS radios and uses the following hexadecimal word: F3A0hex.

The Signal field defines which type of modulation must be used to receive the incoming MPDU. The binary value in this field is equal to the data rate multiplied by 100 kbit/s. In the IEEE 802.11 base standard, two rates are supported: 0Ah for 1 Mbit/s differential binary phase shift keying (DBPSK) and 14hex for 2 Mbit/s differential quadrature phase shift keying (DQPSK).

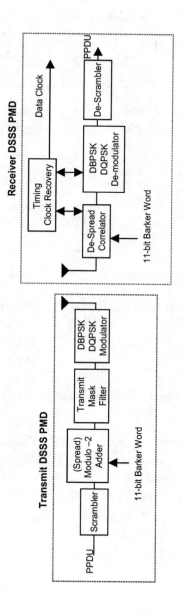

Figure 11–2:   Transmit and receive DSSS PMD

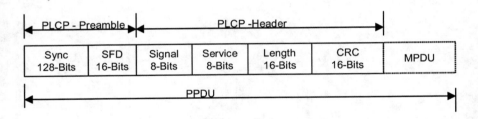

Figure 11–3: DSSS PHY PLCP preamble, PLCP header, and MPDU

The Service field is reserved future use. However, the default value is 00h.

The Length field is an unsigned 16-bit integer that indicates the number of microseconds necessary to transmit the MPDU. The MAC uses this field to determine the end of a PPDU frame.

The CRC field contains the results of a calculated frame check sequence (FCS) from the sending STA. The calculation is performed prior to data scrambling. The CCITT CRC-16 error detection algorithm is used to protect the Signal, Service, and Length fields. The CRC-16 algorithm is represent by the following polynomial: $G(x) = x^{16} + x^{12} + x^5 + 1$. The receiver performs the calculation on the incoming Signal, Service, and Length field values and compares the results against the transmitted value. If an error is detected, the receiver's MAC makes the decision about whether the incoming PPDU should be terminated.

Embedded at the end of the MPDU portion of the PPDU is the FCS field. This field contains a 32-bit cyclic redundancy code (CRC), which protects the information in the PLCP service data unit (PSDU). The DSSS PHY does not determine whether errors are present in the MPDU. The MAC makes that determination similar to the method used by the PHY.

Data scrambling

All information bits transmitted by the DSSS PMD are scrambled using a self-synchronizing 7-bit polynomial. The scrambling polynomial for the DSSS PHY is $G(z) = z^{-7} + z^{-4} + 1$. Scrambling is used to randomize the data in the SYNC field of the PLCP and data patterns that contain long strings of binary ones or zeros. The receiver can descramble the information bits without prior knowledge from the sending station.

DSSS modulation

The DSSS PMD uses differential phase shift keying (DPSK) as the modulation to transmit the PPDU. Two flavors of DPSK are specified. The DSSS PMD transmits the PLCP preamble and PLCP header at 1 Mbit/s using DBPSK. The MPDU is sent at either 1 Mbit/s DBPSK or 2 Mbit/s using DQPSK, depending upon the content in the Signal field of the PLCP header.

DPSK is a modulation technique that uses a balance in-phase/quadrature (I/Q) modulator to generate an RF carrier. The RF carrier is phase modulated and carries symbols mapped from the binary bits in the PPDU. The symbols contain PPDU information. At the receiving STA, data recovery for DPSK is based on the phase differences between two consecutive symbols from the sending STA. DPSK is noncoherent; a clock reference is not needed to recover the data. For 1 Mbit/s DBPSK, 1 and 0 binary bits in the PPDU constitute phase shifts of a 180 degrees, and the signal information is contained on the I arm. For 2 Mbit/s DQPSK, two binary bits are combined from the PPDU to generate the following I/Q symbol pairs: (00,01,11,00). The phase shifts occur at 90 degrees for DQPSK as shown in constellation patterns in Figure 11–4. DQPSK could be thought of as transmitting two 1 Mbit/s DBPSK signals, one on the I and the other on the Q. DBPSK is more tolerant to intersymbol interference (ISI) caused by noise and multipath over the media; therefore, DBPSK is used for the PLCP preamble.

For IEEE 802.11 DSSS Wi-FI products, the rotation of DBPSK- and DQPSK-modulated symbols spinning about the I/Q constellation is *counterclockwise*. This direction is noteworthy because it is common to develop DSSS WLAN products that rotate in the opposite direction. The spinning rotation using DBPSK and DQPSK is illustrated in Figure 11–4.

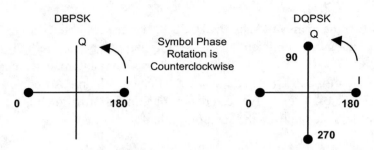

Figure 11–4:   Constellation patterns for DBPSK and DQPSK

Barker spreading method

The DSSS PHY is one of the two 2.4 GHz RF PHYs from which to choose in the IEEE 802.11 standard. Direct sequence is the spreading method used. An 11-bit Barker word is used as the spreading sequence, and every STA in an IEEE 802.11 network uses the same 11-bit sequence. Barker word is classified as short sequences and is known to have very good correlations properties.

Barker word (11 bits) +1, –1, +1, +1, –1, +1, +1, +1, –1, –1, –1

In the transmitter, the 11-bit Barker word is applied to a modulo-2 adder (XOR function) together with each of the information bits in the (scrambled) PPDU as shown in Figure 11–5. The PPDU is clocked at the information rate, e.g., 1 Mbit/s; and the 11-bit Barker word, at 11 Mbit/s (i.e., the chipping clock). The XOR function combines both signals by performing a modulo-2 addition on each PPDU bit along with each bit (sometimes referred to as a *chip*) of the Barker word. The output of the modulo-2 adder results in a signal with at a data rate that is 10 times higher than the information rate. The result in the frequency domain is a signal that is spread over a wider bandwidth at a reduced RF power level. At the receiver, the DSSS signal is convolved with the 11-bit Barker word and correlated. The correlation operation recovers the PPDU information bits at the transmitted information rate, and the undesired interfering in-band signals are spread out of band. The spreading and despreading of narrowband to a wideband signal is commonly referred to as *processing gain* and measured in decibels. Progressing gain is the ratio of the

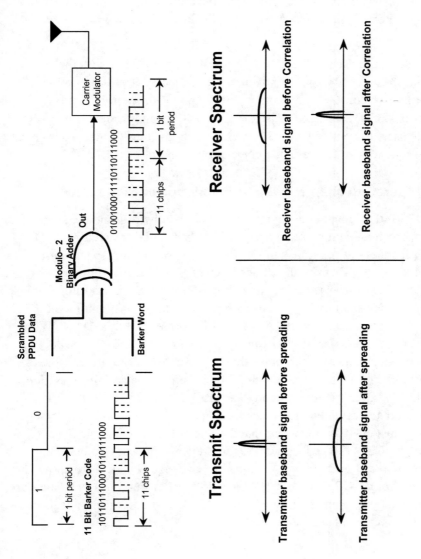

Figure 11–5: DSSS PMD transmitter

DSSS signal rate to the PPDU information rate. In 1999, the FCC made a regulatory rules change under Part 15 that removed the 10 dB processing gain requirement as part of the emissions certification for radios such as IEEE 802.11 Wi-Fi radio products operating in the license-free 2.4 GHz frequency band.

The Barker word used in IEEE 802.11 is not to be confused with the spreading codes used in code division multiple access (CDMA), global positioning system (GPS), and other cellular phone or satellite radio applications. CDMA and GPS use orthogonal spreading codes, which allow multiple users to operate on the same channel frequency. CDMA codes have longer sequences and have richer correlation properties.

DSSS operating channels and transmit power requirements

Each DSSS PHY channel occupies 22 MHz of bandwidth, and the spectral shape of the channel represents a filtered SinX/X function (see Figure 11–6). The DS channel transmit mask in IEEE 802.11 specifies that spectral products be filtered to –30 dBr from the center frequency and all other products be filtered to –50 dBr. This specification allows for three noninterfering channels spaced 25 MHz apart in the 2.4 GHz frequency band. For example, the channel arrangement for North America is illustrated in Figure 11–7. With this channel arrangement, a user can configure multiple DSSS networks to operate simultaneously in the same area. In IEEE 802.11, fourteen center frequency channels are defined with 5 MHz channel spacing for operation across the 2.4 GHz frequency band (see Table 11–1). However, for WLANs operating on adjacent channels, such as channel 6 and channel 7, special attention must be given to ensure that there is proper spacing and distance between the STAs (i.e., clients) or APs to prevent adjacent channel interference (see Figure 11–8). In North America, 12 channels are allowed that range from 2.412 GHz to 2.462 GHz. In most of Europe, 13 channels are allowed that range from 2.412 GHz to 2.472 GHz; and in Japan, one channel frequency is reserved at 2.484 GHz.

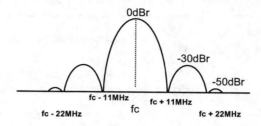

Figure 11–6: Transmit channel shape

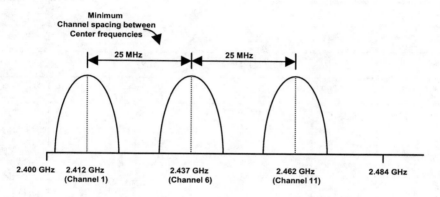

Figure 11–7: Minimum channel spacing for DSSS networks in North America

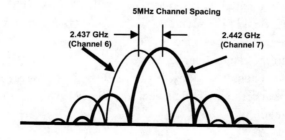

Figure 11–8: Overlapping adjacent channels

Table 11–1: DSSS channels for various parts of the globe

Channel number	Frequency (GHz)	North America	Europe	Spain	France	Japan-MPHPT
1	2.412	X	X			
2	2.417	X	X			
3	2.422	X	X			
4	2.427	X	X			
5	2.432	X	X			
6	2.437	X	X			
7	2.442	X	X			
8	2.447	X	X			
9	2.452	X	X			
10	2.457	X	X	X	X	
11	2.462	X	X	X	X	
12	2.467		X		X	
13	2.472		X		X	
14	2.484					X

In addition to frequency and bandwidth allocations, transmit power is a key parameter that is regulated worldwide. The maximum allowable radiated emission for the DSSS PHY operating in the 2.4 GHz frequency band varies from region to region as illustrated in Table 11–2. The transmit power is directly related to the range a particular IEEE 802.11 DSSS PHY implementation can achieve. Many of the IEEE 802.11 DSSS PHY Wi-Fi products on the market today have selected 100 mW as the nominal RF transmit power level.

Table 11–2: Maximum allowable transmit power worldwide in the 2.4 GHz band

1000 mW	North America
100 mW	Europe
10 mW/MHz	Japan

FREQUENCY HOPPING SPREAD SPECTRUM (FHSS) PHY

As with the DSSS PHY, the FHSS PHY is one of the three PHYs supported in the IEEE 802.11 standard and uses the 2.4 GHz spectrum as the transmission media. Data transmission over the media is controlled by the FHSS PMD sublayer as directed by the FHSS PLCP sublayer. The FHSS PMD takes the binary bits of information from the whitened PSDU and transforms them into RF signals for the wireless media by using carrier modulation and FHSS techniques. Figure 11–9 illustrates the basic of elements of the FHSS PMD transmitter and receiver.

FHSS PLCP sublayer

The PLCP preamble, PLCP header, and PSDU make up the PPDU as shown in Figure 11–10. The PLCP preamble and PLCP header are unique to the FHSS PHY. The PLCP preamble is used to acquire the incoming signal and synchronize the receiver's demodulator. The PLCP header contains information about PSDU from the sending frequency hopping (FH) PHY. The PLCP preamble and PLCP header are transmitted at 1 Mbit/s (i.e., the basic rate).

The SYNC field contains a string of alternating zeros and ones pattern and is used by the receiver to synchronize the receiver's packet timing and to correct for frequency offsets.

The SFD field contains information marking the start of a PSDU frame. A common SFD is specified for all IEEE 802.11 FHSS radios using the following bit pattern: 0000110010111101. The leftmost bit is transmitted first.

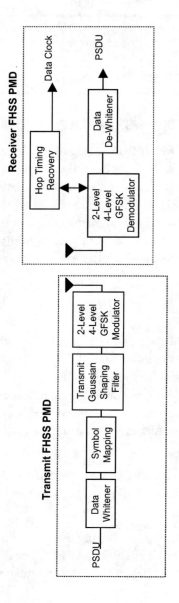

Figure 11–9: Transmit and receive FHSS PMD

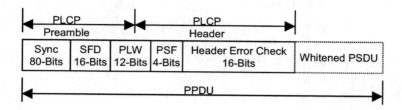

Figure 11–10: FHSS PHY PLCP preamble, PLCP header, and PSDU

The PSDU length word (PLW) field specifies the length of the PSDU in octets and is used by the MAC to detect the end of a PPDU frame.

The PLCP signaling field (PSF) identifies the data rate of the whitened PSDU ranging from 1 Mbit/s to 4.5 Mbit/s increments of 0.5 Mbit/s (see Table 11–3). The PLCP preamble and header are transmitted at the basic rate, i.e., 1 Mbit/s. The optional data rate for the whitened PSDU is 2 Mbit/s.

Table 11–3: PSF bit assignments for PSDU data rates

Bits 1–3	Data rates (Mbit/s)
000	1.0
001	1.5
010	2.0
011	2.5
100	3.0
101	3.5
110	4.0
111	4.5

The Header Check Error field contains the results of a calculated FCS from the sending STA. The calculation is performed prior to data whitening. The

CCITT CRC-16 error detection algorithm is used to protect the PSF and PLW fields. The CRC-16 algorithm is represent by the following polynomial: $G(x) = x^{16} + x^{12} + x^5 + 1$. The receiver performs the calculation on the incoming PSF and PLW fields and compares the results against the transmitted field. If an error is detected, the receiver's MAC determines whether the incoming PPDU should be terminated.

Embedded at the end of the PSDU portion of the PPDU is the FCS field. This field contains a 32-bit CRC, which protects the information in the PSDU. The FHSS PHY does not determine whether errors are present in the PSDU. The MAC makes that determination similar to the method used by the PHY.

PSDU data whitening

Data whitening is applied to the PSDU before transmission to minimize dc bias on the data if long strings of ones or zeros are contained in the PSDU. The PHY stuffs a special symbol every 4 octets of the PDSU in a PPDU frame. A 127-bit sequence generator using the polynomial $[S(x) = x^7 + x^4 + 1]$ and a 32/33 bias-suppression encoding algorithm are used to randomize and whiten the data.

FHSS modulation

The FHSS PMD uses two-level Gaussian frequency shift key (2GFSK) modulation to transmit the PSDU at the basic rate of 1 Mbit/s. The PLCP preamble and PLCP header are always transmitted at 1 Mbit/s. However, four-level GFSK (4GFSK) is supported as an option in the IEEE 802.11 standard that enables the whitened PSDU to be transmitted at a higher rate. The value contained in the PSF field of the PLCP header is used to determine the data rate of the PSDU.

GFSK is a modulation technique used by the FHSS PMD that deviates (i.e., shifts) the frequency on either side of the carrier hopping frequency depending upon whether the binary symbol from the PSDU is either a one or zero. A bandwidth bit period (BT) of 0.5 is used. The changes in the frequency represent symbols containing PSDU information. For 2GFSK, a binary 1 represents the upper deviation frequency from the hopped carrier, and a binary 0 represents the lower deviation frequency. The deviation frequency shall be greater than 110 KHz for IEEE 802.11 FHSS radios. The carrier frequency deviation is given by the following equations:

$$\text{Binary } 1 = F_c + f_d \quad \text{Carrier hopping frequency plus the upper deviation frequency}$$

$$\text{Binary } 0 = F_c - f_d \quad \text{Carrier hopping frequency minus the lower deviation frequency}$$

The 4GFSK is similar to 2GFSK and is used to achieve a data rate of 2 Mbit/s in the same occupied frequency bandwidth. The modulator combines two binary bits from the whitened PSDU and encodes them into the following symbol pairs: (10,11,01,00). The symbol pairs generate four frequency deviations from the hopped carrier frequency: two upper and two lower. The symbol pairs are transmitted at 1 Mbit/s; and for each bit sent, the resulting data rate is 2 Mbit/s.

FHSS channel hopping

A set of hopping sequences is defined in IEEE 802.11 for use in the 2.4 GHz frequency band. The channels are evenly spaced across the band over a span of 83.5 MHz. During the development of the IEEE 802.11 standard, the hopping sequences listed in the standard were preapproved for operation in North America, Europe, and Japan. The required number of hopping channels is depended upon the geographic location. In North America and in Europe (excluding Spain and France), the number of hopping channels is 79. The number of hopping channels for Spain is 23; and for France, 35. In Japan, the required number of hopping channels is 23. The hopping center channels are spaced uniformly across the 2.4 GHz frequency band occupying a bandwidth of 1 MHz. In North America and in Europe (excluding Spain and France), the hopping channels operate from 2.402 GHz to 2.480 GHz; and for Japan, 2.473 GHz to 2.495 GHz. In Spain, the hopping channels operate from 2.447 GHz to 2.473 GHz; and for France, 2.448 GHz to 2.482GHz. Channel 2 is the first hopping channel located at a center frequency of 2.402 GHz, and channel 95 is the last hopping frequency channel in the 2.4 Ghz band centered at 2.495 Ghz.

Channel hopping is controlled by the FHSS PMD. The FHSS PMD transmits the whitened PSDU by hopping from channel to channel in a pseudo-random fashion using one of the hopping sequences. The hop rate is set by the regulatory bodies in the country of operation. In the United States, FHSS radios must hop a minimum of 2.5 hops per second for a minimum hop distance of 6 MHz to be in accordance with the rules specified by the FCC rules under Part 15.

The hopping sequences for IEEE 802.11 are grouped in hopping sets, i.e., Set 1, Set 2, and Set 3, for worldwide operation. The sequences are selected when a FHSS BSS is configured for a WLAN. The hopping sets are design to minimize interference between neighboring FHSS radios in a set. The following hopping sets are valid IEEE 802.11 hopping sequence numbers:

Operation in North America and most of Europe:

Set 1: (0, 3, 6, 9, 12, 15, 18, 21, 24, 27, 30, 33, 36, 39, 42, 45, 48, 51, 54, 57, 60, 63, 66, 69, 72, 75)

Set 2: (1, 4, 7, 10, 13, 16, 19, 22, 25, 28, 31, 34, 37, 40, 43, 46, 49, 52, 55, 58, 61, 64, 67, 70, 73, 76)

Set 3: (2, 5, 8, 11, 14, 17, 20, 23, 26, 29, 32, 35, 38, 41, 44, 47, 50, 53, 56, 59, 62, 65, 68, 72, 74, 77)

Operation in Spain:

Set 1: (0, 3, 6, 9, 12, 15, 18, 21, 24)

Set 2: (1, 4, 7, 10, 13, 16, 19, 22, 25)

Set 3: (2, 5, 8, 11, 14, 17, 20, 23, 26)

Operation in France:

Set 1: (0, 3, 6, 9, 12, 15, 18, 21, 24, 27, 30)

Set 2: (1, 4, 7, 10, 13, 16, 19, 22, 25, 28, 31)

Set 3: (2, 5, 8, 11, 14, 17, 20, 23, 26, 29, 32)

Operation in Japan:

Set 1: (6, 9, 12, 15)

Set 2: (7, 10, 13, 16)

Set 3: (8, 11, 14, 17)

INFRARED (IR) PHY

The IR PHY is one of the three initial PHYs supported in the IEEE 802.11 standard when ratified in 1997. The IR PHY differs from DSSS and FHSS because IR uses near-visible light as the transmission media. IR communications relies on light energy, which is reflected off objects or by line of sight. The IR PHY operation is restricted to indoor environments and cannot pass through walls, as can DSSS and FHSS radio signals. Data transmission over the media is controlled by the IR PMD sublayer as directed by the IR PLCP sublayer. The IR PMD takes the binary bits of information from the PSDU and transforms them into light energy emissions for the wireless media by using carrier modulation. Figure 11–11 illustrates the basic of elements of the FHSS PMD transmitter and receiver.

IR PLCP sublayer

The PLCP preamble, PLCP header, and PSDU make up the PPDU as shown in Figure 11–12. The PLCP preamble and PLCP header are unique to the IR PHY. The PLCP preamble is used to acquire the incoming signal and synchronize the receiver prior to the arrival of the PSDU. The PLCP header contains information about PSDU from the sending IR PHY. The PLCP preamble and PLCP header are always transmitted at 1 Mbit/s, and the PSDU can be sent at either 1 Mbit/s or 2 Mbit/s.

The SYNC field contains a sequence of alternated presence and absence of a pulse in consecutive time slots. The SYNC field is used by the IR PHY to perform signal acquisition and clock recovery. The IEEE 802.11 standard specifies 57 time slots as the minimum and 73 time slots as the maximum.

The SFD field contains information that marks the start of a PPDU frame. A common SFD is specified for all IEEE 802.11 IR implementations. The SFD is represented by the following bit pattern: 1001.

The Data Rate field defines the data rate at which the PPDU is transmitted. There are two rates from which to choose: 000 for 1 Mbit/s (i.e., the basic rate) and 001 for 2 Mbit/s (i.e., the enhanced access rate). The PLCP preamble and PLCP header are always sent at the basic rate, i.e., 1 Mbit/s.

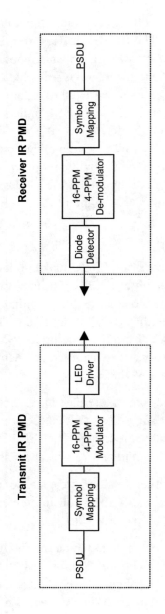

Figure 11–11:   Transmit and receive IR PMD

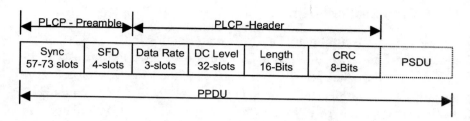

Figure 11–12: IR PHY PLCP preamble, PLCP header, and PSDU

The DC Level field contains information that allows the IR PHY to stabilize the dc level after receiving the preamble and data rate fields. The supported data rates use the following bit patterns:

1 Mbit/s: 00000000100000000000000010000000

2 Mbit/s: 00100010001000100010001000100010

The Length field contains an unsigned 16-bit integer that indicates the number of microseconds to transmit the PSDU. The MAC layer uses this field to detect the end of a frame.

The CRC field contains the calculated 16-bit CRC result from the sending station. The CCITT CRC-16 error detection algorithm is used to protect the Length field. The CRC-16 algorithm is represent by the following polynomial: $G(x) = x^{16} + x^{12} + x^5 + 1$. The receiver performs the calculation on the incoming Length field and compares the results against the transmitted field. If an error is detected, the receiver's MAC determines whether the incoming PSDU should be terminated.

Embedded at the end of the PSDU portion of the PPDU is the FCS field. This field contains a 32-bit CRC, which protects the information in the PSDU. The IR PHY does not determine whether errors are present in the PSDU. The MAC makes that determination similar to the method used by the PHY.

IR PHY modulation method

The IR PHY transmits binary data at 1 Mbit/s and 2 Mbit/s using a modulation known as pulse position modulation (PPM). PPM is used in IR systems to reduce the optical power required of the LED IR source. The specific data rate is dependent upon the type of PPM. The modulation for 1 Mbit/s operation is 16-PPM; and for 2 Mbit/s, 4-PPM. PPM is a modulation technique that keeps the amplitude and pulse width constant and varies the position of the pulse in time. Each position represents a different symbol in time.

For 16-PPM, each group of data bits of the PSDU is mapped to one of the 16-PPM symbols for 1 Mbit/s operation. Notice in Table 11–4 a "1" bit in the 16-PPM symbol column represents data bit position. The order of the transmission bits is left to right. The data bits are arranged (i.e., gray coded) to reduce the possibility of multiple bit errors due to intersymbol interference in the media.

Table 11–4:   16-PPM symbol map for 1 Mbit/s

Data bits	16-PPM symbols
0000	0000000000000001
0001	0000000000000010
0011	0000000000001000
0010	0000000000010000
0110	0000000000100000
0111	0000000001000000
0101	0000000010000000
0100	0000000100000000
1100	0000001000000000
1101	0000010000000000
1111	0000100000000000

Table 11–4: 16-PPM symbol map for 1 Mbit/s *(Continued)*

Data bits	16-PPM symbols
1110	0000100000000000
1010	0001000000000000
1011	0010000000000000
1001	0100000000000000
1000	1000000000000000

For 2 Mbit/s operation, 4-PPM is used; and two data bits are paired in the PSDU to form a 4-bit symbol map as shown in Table 11–5. The transmission order of the bits is left to right.

Table 11–5: 4-PPM symbol map for 2 Mbit/s

Data bits	4-PPM symbol
00	0001
01	0010
11	0100
10	1000

GEOGRAPHIC REGULATORY BODIES

IEEE 802.11 WLAN DSSS, FHSS, complementary code keying (CCK), and OFDM radios operating in the 2.4 GHz frequency band and OFDM operating in the 5 GHz frequency band must comply with the local geographical regulatory domains before operating in this spectrum. These products are subject to certification. The technical requirements in the IEEE 802.11 standard were developed to comply with the regulatory agencies in North America, Australia, Europe, and Japan. The regulatory agencies in these regions set emission requirements for WLANs to minimize the amount of interference a radio can generate or receive from another in the same proximity. The regulatory requirements do not affect the interoperability of IEEE 802.11 Wi-Fi products. It is the responsibility of the product developers to check with the regulatory agencies for specific rules and regulations for compliance. In some cases, additional certifications are necessary for regions within Europe or outside of Japan or North America and in other continents. Listed below are some agencies discussed in IEEE 802.11.

North America

> Approval Standards: Industry Canada
>
> Documents: GL36
>
> Approval Authorities: Federal Communications Commission (FCC)
>
> USA Documents: CFR 47, Part 15 Sections 15.205, 15.209, 15.247
>
> Approval Authority: Industry Canada, FCC (USA)

Spain

> Approval Standards: Supplemento Del Numero 164 Del Boletin Oficial Del Estado (Published 10, July 91, Revised 25 June 93)
>
> Documents: ETS 300-328, ETS 300-339
>
> Approval Authority: Cuadro Nacional De Atribucion De Frecuesias

Europe

> Approval Standards: European Telecommunications Standards Institute
>
> Documents: ETS 300-328, ETS 300-339
>
> Approval Authority: National Type Approval Authorities

Chapter 12 PHY extensions to IEEE 802.11

In 1999, the IEEE ratified two amendments for higher rate PHY extensions to IEEE 802.11. The first extension, IEEE 802.11a, defines requirements for a PHY operating in the 5.0 GHz U-NII frequencies and 5 GHz spectrum worldwide for data rates ranging from 6 Mbit/s to 54 Mbit/s. The second extension, IEEE 802.11b, defines a set of PHY specifications operating in the 2.4 GHz industrial, scientific, and medical (ISM) frequency band for data rates up to 11 Mbit/s. Both PHYs are defined to operate with the existing MAC. Today, many of the Wi-Fi products in the market support both IEEE 802.11a and IEEE 802.11b extensions, and dual band combinations come in variety of platforms for the home consumer, small office, and enterprise applications including Wi-Fi hotspots. This chapter gives the reader a general overview of some of the requirements specified for each extension.

IEEE 802.11a: ORTHOGONAL FREQUENCY DIVISION MULTIPLEXING (OFDM) PHY

The IEEE 802.11a PHY is one of the PHY extensions of IEEE 802.11 and is referred to as the OFDM PHY. The OFDM PHY provides the capability to transmit PSDU frames at multiple data rates up to 54 Mbit/s for WLANs operating in the 5 GHz frequency band where transmission of standard definition television (SDTV), high-definition television (HDTV), multimedia content, dense deployment of STAs (i.e., clients) per AP, or less interference is a consideration. The OFDM PHY defined for IEEE 802.11a is similar to the OFDM PHY specification of ETSI HIPERLAN II. In late 2002, IEEE 802.11a was chosen as the standard for adoption for operation in the 5 GHz frequency band worldwide.

In the OSI structure, the PHY's PLCP sublayer and PMD sublayer are unique to the OFDM PHY. The following sections give an overview of the PLCP header, data rates, and modulations defined in IEEE 802.11a.

OFDM PLCP sublayer

The PPDU is unique to the OFDM PHY. The PPDU frame consists of a PLCP preamble, Signal field, and Data field as shown in Figure 12–1. The receiver uses the PLCP preamble to acquire the incoming OFDM signal and synchronize the demodulator. The PLCP header (found in the Signal field) contains information about the PSDU from the sending OFDM PHY. The PLCP preamble and PLCP header are always transmitted at 6 Mbit/s, BPSK-OFDM-modulated using convolutional encoding rate R = 1/2.

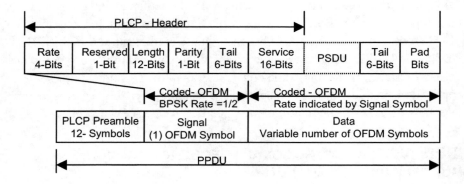

Figure 12–1:   OFDM PLCP preamble, PLCP header, and PSDU

The PLCP preamble is used to acquire the incoming signal and to train and synchronize the receiver. The PLCP preamble consists of 12 symbols, ten of which are short symbols and two are long symbols. The preamble duration is 16 μs, and the time is equally divided between the short and long symbols. The short symbols are used to train the receiver's automatic gain control (AGC) and to obtain a coarse estimate of the channel. The long symbols are used to finetune the channel estimates. AGC and coarse channel estimates and the long symbols are used to finetune channel offsets. Twelve subcarriers are used for the short symbols; and 53 subcarriers, for the long symbols. The training of an OFDM is accomplished in 16 μs. PLCP preamble is BPSK-OFDM-modulated at 6 Mbit/s.

The Signal field is 24 bits long and contains information about the rate and length of the PSDU. The Signal field is BPSK-OFDM-modulated using convolutional encoding rate R = 1/2,. Four bits (R1–R4) are used to encode the rate. Eleven bits are defined for the length, and the remaining bits are defined for parity and signal tail. The rate bits (R1–R4) are defined in Table 12–1. The mandatory data rates for IEEE 802.11a systems are 6 Mbit/s through 24 Mbit/s; however, Wi-Fi devices must support the highest rate, 54 Mbit/s, for certification.

Table 12–1:   PSDU data rate selection

Rate (Mbit/s)	Modulation	Coding rate	Signal bits (R1-R4)
6	BPSK	R = 1/2	1101
9	BPSK	R = 3/4	1111
12	QPSK	R = 1/2	0101
18	QPSK	R = 3/4	0111
24	16-QAM	R = 1/2	1001
36 (optional)	16-QAM	R = 3/4	1011
48 (optional)	64-QAM	R = 2/3	0001
54 (optional)	64-QAM	R = 3/4	0011

Note: For Wi-Fi certification, the following data rates are mandatory: 6, 9, 12, 18, 24, and 54 Mbit/s.

The Length field is an unsigned 12-bit integer that indicates the number of octets in the PSDU.

The Data field contains the Service subfield, PSDU, tails bits, and pad bits. A total of six tail bits containing zeros are appended to the PPDU to ensure that the convolutional encoder is brought back to zero state. The equation for determining the number of bits in the Data field, the number of tail bits, the number of OFDM symbols, and the number pad bits is defined in IEEE 802.11a.

IEEE 802.11a modulation

The preshared key (PSK) and quadrature amplitude modulation (QAM) are two types digital modulation used for IEEE 802.11a. At the lower data rates, binary phase shift keying (BPSK) and quadrature phase shift keying (QPSK) are used; and at the higher data rates of 48 Mbit/s to 54 Mbit/s, 64-QAM is used. The preamble is always BPSK-modulated at 6 Mbit/s. The amplitude and phase for each subcarrier are independent of each other; therefore, for 16-QAM, 16 points are generated in a constellation; and for 64-QAM, 64 points (see Figure 12–2).

PLCP and data scrambler

All information bits transmitted by the OFDM PMD are scrambled using a self-synchronizing 127-bit sequence generator. The scrambling polynomial for the OFDM PHY is $S(x) = x^{-7} + x^{-4} + 1$. Scrambling is used to randomize the Service field, PSDU, tail bits, pad bits and data patterns, which contain long strings of binary ones or zeros. Prior to sending a PPDU frame, the seven LSBs of the Service field are reset to zero prior to data scrambling in order to estimate the state of the scrambler. The receiver can descramble the information bits without prior knowledge from the sending STA.

Convolutional encoding

Convolutional encoding is usually applied to OFDM radios. This coding technique is a form of forward error-correction. All information contained in the Service field, PSDU, tail bits, and pad bits are encoded using convolutional encoding rate R = 1/2, 2/3, or 3/4 corresponding to the desired data rate. The lower data rate of 6 Mbit/s uses R = 1/2, and 54 Mbit/s uses R = 3/4. Convolutional encoding is generated using the following polynomials: $g_0 = 133_8$ and $g_1 = 171_8$ of rate R = 1/2. Puncture codes are used for the higher data rates. Industry standard algorithms such as Viterbi are recommended for encoding and decoding.

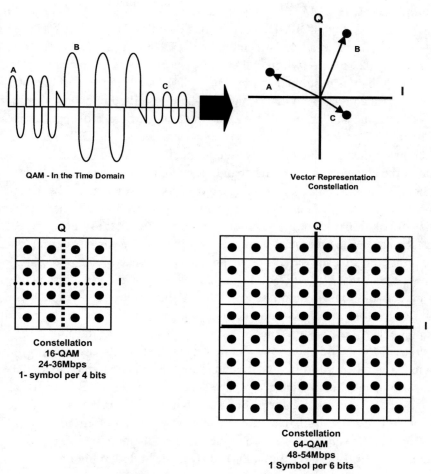

Figure 12–2:   QAM modulation

IEEE 802.11a OFDM

As part of the IEEE ratification of IEEE 802.11a in 1999, the IEEE 802.11 Working Group adopted OFDM modulation as the basis for that amendment. The OFDM method chosen is similar to the modulation technique adopted in Europe by ETSI HIPERLAN II 5 GHz radio PHY specification. OFMD has been around for many decades and has been used in many code division multiple access (CDMA) transmission systems. Because of its unique benefits and robustness, today OFMD is being considered as the baseline for many of the IEEE 802 wireless standards including IEEE 802.16a™ (wireless MAN) and IEEE 802.15-3a™ (wireless PAN ultra wideband). It is also used in IEEE 802.11g, which is described in Chapter 14. The basic principal of operation first divides a high-speed binary signal to be transmitted into a number of lower data rate subcarriers. There are 48 data subcarriers and 4 carrier pilot subcarriers for a total of 52 nonzero subcarriers defined in IEEE 802.11a. Each lower data rate bit stream is used to modulate a separate subcarrier from one of the channels in the 5 GHz band. Intersymbol interference (ISI) is generally not a concern for lower speed carriers; however, the subchannels may be subjected to frequency selective fading. Therefore, bit interleaving and convolutional encoding are used to improve the bit error rate performance. The scheme uses integer multiples of the first subcarrier, which are orthogonal to each other. This technique is known as *orthogonal frequency division multiplexing* (OFDM). Prior to transmission, the PPDU is encoded using convolutional coding rate R = 1/2, and the bits are reorder and bit-interleaved for the desired data rate. The value R signifies the coding rate. For example, R = 1/2 transmits one bit for every three bits. As the code rate decreases, the encoding becomes more robust. Each bit is then mapped into a complex number according the modulation type and subdivided in 48 data subcarriers and 4 pilot subcarriers. The subcarriers are combined using an inverse Fast Fourier Transform (FFT) and transmitted. At the receiver, the carrier is converted back to a multicarrier lower data rate form using an FFT. The lower data subcarriers are combined to form the high rate PPDU. An example of an IEEE 802.11a OFDM PMD is illustrated in Figure 12–3.

OFDM operating channels and regulatory domains

The 5 GHz frequency band is segmented into four contiguous bands for worldwide operation. Each of the bands has a fixed number of channels and limits on transmit RF power. The lower band ranges from 5.15 GHz to 5.25 GHz, the middle band ranges from 5.25 GHz to 5.35 GHz, and the upper band ranges from 5.725 GHz to 5.825 GHz. In 2003, the FCC adopted a notice of proposed rulemaking allocating an additional 255 MHz of spectrum to the middle U-NII band ranging from 5.470 GHz to 5.725 GHz for users

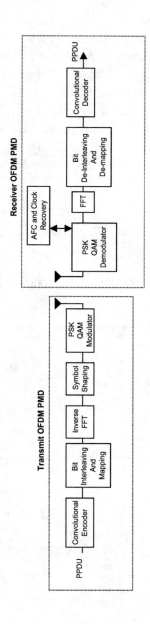

Figure 12–3:   IEEE 802.11a transmit and receive OFDM PMD

operating in the United States. This additional spectrum provides 23 channels over a total bandwidth of 555 MHz for IEEE 802.11a Wi-Fi devices operating from 5.150 GHz to 5.825 GHz. The existing lower and middle bands accommodate eight channels in a total bandwidth of 200 MHz, and the upper band accommodates four channels in a 100 MHz bandwidth. The channel center frequencies are spaced 20 MHz apart. However, the outer most channels of the lower and middle bands are centered 30 MHz from the outer edges. The upper band is centered 20 MHz from the outer edges. The channel frequencies and numbers are defined in IEEE 802.11a in 5 MHz increments starting at 5 GHz (see Figure 12–4). A set of channel frequencies for each of the frequencies is defined in Table 12–2.

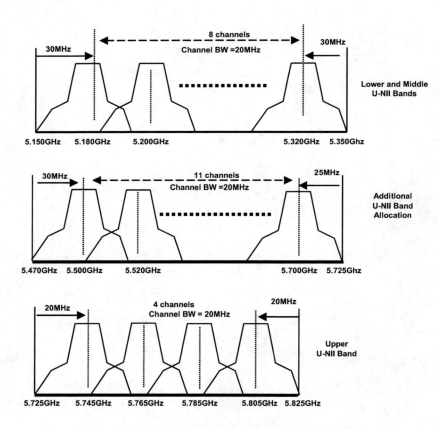

Figure 12–4:   5 GHz frequency band channels of operation

Table 12–2: Channel frequencies and channel numbers for the 5 GHz band worldwide

Regulatory domain	Frequency band (GHz)	Channel number	Center frequencies (GHz)
United States Canada Europe Japan China	U-NII lower band 5.150–5.250	36 40 44 48	5.180 5.200 5.220 5.240
United States Canada Europe China	U-NII middle band 5.250–5.350	52 56 60 64	5.260 5.280 5.300 5.320
United States Canada Europe China	Additional U-NII band 5.470–5.725	100 104 108 112 116 120 124 128 132 136 140	5.500 5.520 5.540 5.560 5.580 5.600 5.620 5.640 5.640 5.680 5.700
United States	U-NII upper band 5.725–5.825	149 153 157 161	5.745 5.765 5.785 5.805

Transmit power requirements

Each of the IEEE 802.11a OFDM channels occupies 20 dB of bandwidth and conforms to a spectral mask as shown in Figure 12–5. The transmit mask is specified with a filtered shape to –40 dBr at 30 MHz from the center frequency to allow operation of overlapping channels with minimal adjacent channel interference.

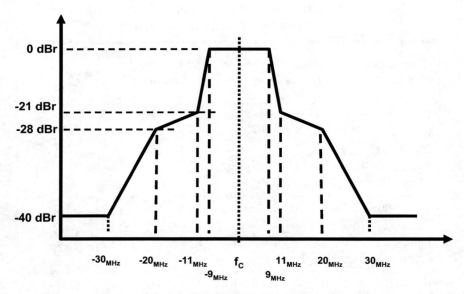

Figure 12–5: IEEE 802.11a transmit mask

In addition to frequency and channel allocations, transmit power is a key parameter regulated in the 5 GHz U-NII frequency band. Four transmit RF power levels are specified: 40 mW, 200 mW, 800 mW, and 1000 mW as illustrated in Table 12–3. The upper band defines RF transmit power levels suitable for bridging applications while the lower band specifies a transmit power level suitable for short-range indoor home and small office. The middle band is suitable for Wi-Fi hotspots and enterprise environments.

**Table 12–3: Transmit power levels
for North American operation**

Frequency band (GHz)	Maximum output transmit power with 6 dBi antenna gain
5.150–5.250	40 mW (2.5 mW/MHz)
5.250–5.350	200 mW (12.5 mW/MHz)

**Table 12–3: Transmit power levels
for North American operation *(Continued)***

Frequency band (GHz)	Maximum output transmit power with 6 dBi antenna gain
5.470–5.725	1000 mW
5.725–5.825	800 mW (50 mW/MHz)

Geographic regulatory bodies

IEEE 802.11a WLAN OFDM Wi-Fi radios operating in the 5 GHz frequency band must comply with the local geographical regulatory domains before operating in this spectrum. These products are subject to certification. At the time IEEE 802.11a was being developed, the technical requirements were specified to comply with the regulatory agencies in North America, Australia, Europe, and Japan. The regulatory agencies in these regions set emission requirements for WLANs to minimize the amount of interference a radio can generate or receive from another in the same proximity. The regulatory requirements do not affect the interoperability of IEEE 802.11a Wi-Fi products. It is the responsibility of the product developers to check with the regulatory agencies for the necessary certifications. In the United States, the FCC is responsible for the allocation of 5 GHz frequency bands. In Canada, the Industry Canada is responsible; in Spain, Cuadro Nacional; and in Europe, ETSI. Listed below are details on the regulatory agencies and reference documents recommended by IEEE 802.11.

North America

Approval Standards: Industry Canada

Documents: GL36

Approval Authorities: Federal Communications Commission, (FCC)

USA Documents: CFR 47, Part 15 Sections 15.205, 15.209, 15.247

Approval Authority: Industry Canada, FCC (USA)

Spain

> Approval Standards: Supplemento Del Numero 164 Del Boletin Oficial Del Estado (Published 10, July 91, Revised 25 June 93)
>
> Documents: ETS 300-328, ETS 300-339
>
> Approval Authority: Cuadro Nacional De Atribucion De Frecuesias

Europe

> Approval Standards: European Telecommunications Standards Institute (ETSI) CEPT
>
> Documents: ETS 300-328, ETS 300-339, ETSI EN 301893
>
> Approval Authority: National Type Approval Authorities

Globalization of spectrum at 5 GHz

In 2001, Japan's Multimedia Mobile Access Communication System (MMACS) adopted the use of IEEE 802.11a WLANs operating in the 5.150–5.250 GHz frequency band as the primary choice of spectrum. In 2003, the World Radiocommunication Conference (WRC 2003) adopted the harmonization and globalization of the 5 GHz frequency band from 5.150 GHZ to 5.725 GHz allowing the use of IEEE 802.11a devices in Europe and worldwide, providing that the mechanisms for transmit power control (TPC) and dynamic frequency selection (DFS) specified in IEEE 802.11h are implemented in STA (i.e., client) devices and APs. Satellites and radars remain the primary users of the 5 GHz frequency band. Therefore, IEEE 802.11a devices operating in Europe must employ radar detection mechanisms as described in IEEE 802.11h and must be capable of stopping RF emissions and transmission of packet data in the presence of the primary users, e.g., radars. The radar detection and TPC mechanisms for IEEE 802.11h are described in Chapter 6. At the time, we were writing this book, the FCC was in the final stages of issuing a report and order on a decision mandating a requirement to use the mechanisms in IEEE 802.11h for IEEE 802.11a devices operating in the 5.470–5.725 GHz frequency band.

IEEE 802.11b: 2.4 HIGH-RATE DIRECT SEQUENCE SPREAD SPECTRUM (HR/DSSS) PHY

The IEEE 802.11b PHY is one of the PHY extensions of IEEE 802.11 and is referred to as HR/DSSS PHY. The HR/DSSS PHY provides two functions. First, the HR/DSSS extends the PSDU data rates to 5.5 Mbit/s and 11 Mbit/s using an enhanced modulation technique. Second, the HR/DSSS PHY provides a rate shift mechanism, which allows 11 Mbit/s networks to fall back to 1 Mbit/s and 2 Mbit/s and to interoperate with the legacy IEEE 802.11 2.4 GHz RF PHYs. The OSI structure and operation of the PHY's PLCP sublayer and PMD sublayer for HR/DSSS is similar to the existing IEEE 802.11 DSSS PHY described in Chapter 11. The following sections give an overview of the PLCP header, data rates, and modulations defined in IEEE 802.11b.

HR/DSSS PLCP sublayer

A PPDU frame consists of the PLCP preamble, PLCP header, and the PSDU. As with IEEE 802.11 DSSS, the PMD uses the PLCP preamble to acquire the incoming signal and synchronize the receiver's demodulator. The HR/DSSS PHY defines two PLCP preambles: long and short (see Figure 12–6). The long preamble uses the same PLCP preamble and PLCP header as the IEEE 802.11 DSSS PHY and sends the information at 1 Mbit/s using DBPSK modulation and Barker word direct sequence spreading. The PSDU is transmitted at 1, 2, 5.5, and 11 Mbit/s as determined by the content in the Signal field. The long preamble is backwards compatible with existing IEEE 802.11 DSSS PHYs and defined to interoperate with existing IEEE 802.11 Wi-Fi WLANs operating at 1 Mbit/s and 2 Mbit/s.

The short preamble uses a 56-bit SYNC field to acquire the incoming signal and transmits the preamble at 1 Mbit/s using DBPSK modulation and Barker word spreading. The PLCP header transmits at 2 Mbit/s using DQPSK modulation and Barker word spreading (see Figure 12–6). In this case, the PSDU is transmitted at 2, 5.5, or 11 Mbit/s as determined by the content in the Signal field. The short preamble is an option in IEEE 802.11b and is useful for networks where throughput efficiency and interoperability with existing IEEE 802.11 DSSS radios is not necessary. There is one caveat: the short preamble

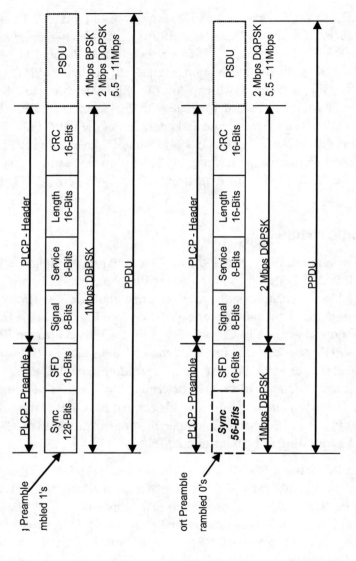

Figure 12–6:   HR/DSSS PHY PLCP long and short preambles, PLCP header, and PSDU

radio can interoperate only with itself. Therefore, for a short preamble radio to be IEEE 802.11b compliant, it must support the long preamble.

The SYNC field is used by the receiver to acquire the incoming signal and synchronize the receiver's carrier tracking and timing prior to receiving the SFD. The long preamble SYNC field is 128 bits in length and contains a string of scrambled ones. The scrambler seed bit pattern used to initialize the scrambler for the long preamble is 01101100. The short preamble SYNC field is 56 bits in length and contains a string of scrambled zeros. The scrambler seed bit pattern used to initialize the scrambler for short preamble operation is 00011010. The short preamble SYNC field is used for networks where minimizing overhead and maximizing PSDU throughput are considerations.

The SFD field contains information marking the start of a PPDU frame. The SFD specified is common for all IEEE 802.11 DSSS and IEEE 802.11b short and long preamble radios. The following hexadecimal word is used: F3A0hex transmitted LSB first.

The Signal field defines which type of modulation must be used to receive the incoming PSDU. The binary value in this field is equal to the data rate multiplied by 100 kbit/s. The 1 Mbit/s data rate is used for long and short preamble implementations. The bit patterns in this field always represent the data rates in Table 12–4.

Table 12–4:   Bit patterns in Signal field

Signal field	Data rate (Mbit/s)
00001010	1 (long preamble only)
00010100	2
00111110	5.5
01101110	11

The Service field uses 3 bits of the reserved 8 bits for IEEE 802.11b. Data bit (bit 2) determines whether the transmit frequency and symbol clocks use the same local oscillator. Data bit (bit 3) indicates whether complementary code keying (CCK) or packet binary convolutional code (PBCC) is used, and data bit (bit 7) is a bit extension used in conjunction with the Length field to calculate the duration of the PSDU, in microseconds. This field is used for the long and short preamble frames.

The Length field is an unsigned 16-bit integer that indicates the number of microseconds necessary to transmit the PSDU. For any data rate over 8 Mbit/s, bit 7 of the Service field is used with the Length field to determine the time, in microseconds, from the number of octets contain in Length field. A calculation is defined in IEEE 802.11b for determining the length, in microseconds, for CCK and PBCC as applied to both preambles. The MAC uses this field to determine the end of a PPDU frame.

The CRC field contains the results of a calculated FCS from the sending STA. The calculation is performed prior to data scrambling for the long and short preamble. The CCITT CRC-16 error detection algorithm is used to protect the Signal, Service, and Length fields. The CRC-16 algorithm is represented by the following polynomial: $G(x) = x^{16} + x^{12} + x^5 + 1$. The receiver performs the calculation on the incoming Signal, Service, and Length field values and compares the results against the transmitted value. If an error is detected, the receiver's MAC makes the decision whether incoming PSDU should be terminated.

High-rate data scrambling

All information bits transmitted by the DSSS PMD are scrambled using a self-synchronizing 7-bit polynomial. The scrambling polynomial for the DSSS PHY is $G(z) = z^{-7} + z^{-4} + 1$. Scrambling is used to randomize the long and short preamble data in the SYNC field of the PLCP and for data patterns that contain long strings of binary ones or zeros. The receiver can descramble the information bits without prior knowledge from the sending STA. The scrambler initialization bit patterns are represented as (00011010) for the short preamble and (01101100) for the long preamble.

IEEE 802.11 high-rate operating channels

The HR/DSSS PHY uses the same frequency channels as defined in Chapter 11 for the IEEE 802.11 DSSS PHY. The channel center frequencies are spaced 25 MHz apart to allow multiple WLANs to operate simultaneous in the same area without interfering each other. An example of a typical channel arrangement for noninterfering channels for North America is illustrated in Figure 12–7.

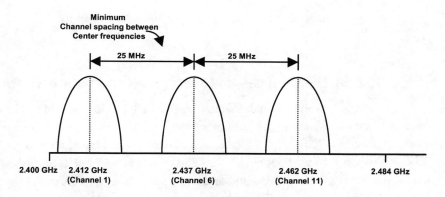

**Figure 12–7:   Minimum channel spacing
for IEEE 802.11 high-rate networks**

In Europe, channel 1 (2.412 GHz), channel 7 (2.442 GHz), and channel 13 (2.472 GHz) are used to form three noninterfering networks. However, for IEEE 802.11b WLANs operating on adjacent channels, such as channel 6 and channel 7, special attention must be given to ensure that there is proper spacing and distance between the STAs (i.e., clients) or APs to prevent adjacent channel interference (see Figure 12–8).

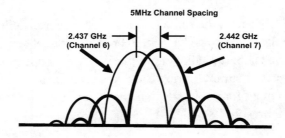

Figure 12–8:   Overlapping adjacent channels

IEEE 802.11 DSSS HIGH-RATE MODULATION AND DATA RATES

Four modulation formats and data rates are defined in IEEE 802.11b. The data rates include the basic rate, the extended rate, and enhanced rate. The basic rate is defined as 1 Mbit/s modulated with DBPSK, and the extended rate is 2 Mbit/s modulated with DQPSK. The 11-bit Barker word is used as the spreading format for the basic rate and extended rate as described for the DSSS PHY in Chapter 11. The enhanced rate is defined to operate at 5.5 Mbit/s and 11 Mbit/s using CCK modulation and PBCC. PBCC is an option in the standard and for networks requiring enhanced performance. Frequency agility is another option defined in IEEE 802.11b. As with the 1 Mbit/s and 2 Mbit/s DSSS PHY, this option enables existing IEEE 802.11 FHSS 1 Mbit/s networks to be interoperable with 11 Mbit/s CCK high-rate networks. The PBCC and frequency agility option are described later in this chapter.

Complementary code keying (CCK) modulation

In July of 1998, the IEEE 802.11 Working Group adopted CCK as the basis for the high-rate extension to deliver PSDU frames at speeds of 5.5 Mbit/s and 11 Mbit/s. CCK was adopted because it easily provides a path for interoperability with existing IEEE 802.11 1Mbit/s and 2 Mbit/s systems by maintaining the same bandwidth and incorporating the existing DSSS PHY PLCP preamble and PLCP header.

CCK is a variation on M-ary Orthogonal Keying modulation and is based on an in-phase (I) and quadrature (Q) architecture using complex symbols. CCK allows for multichannel operation in the 2.4 GHz band by using the existing 1 Mbit/s and 2 Mbit/s DSSS channelization scheme. CCK uses eight complex chips in each spreading code word. Each chip can assume one of four phases (QPSK). CCK uses 64 base spreading code words out of a possible set of 65 536 (i.e., 65 536 – 4^3). Base spreading codes were chosen with good autocorrelation and cross-correlation properties. Each spreading code is 16 bits in length. The CCK modulator chooses one of M unique for transmission of the scrambled PSDU. CCK uses one vector from a set of 64 complex QPSK vectors for the symbol and thereby modulates 6 bits (i.e., one of 64) on each spreading code symbol as shown in Figure 12–9.

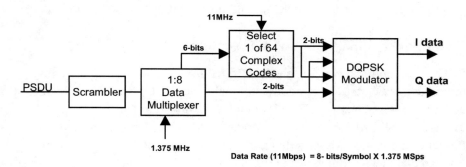

Figure 12–9:   Generation of CCK modulation

Each spreading code is 8 complex chips in length. CCK uses a complex set of Walsh/Hadamard functions known as *complementary codes*. Refer to IEEE 802.11b for the equation used to derive the set of code words. There are four phase terms in the CCK equation. One of the terms modulates all of the chips and is used for the QPSK rotation of the entire code vector. The others modulate every odd chip, every odd pair of chips, and every odd quad of chips. To minimize dc offsets, the fourth and seventh terms in the equation are rotated by 180 degrees with a cover sequence. As with the IEEE 802.11 DSSS PHY, the phase rotation for CCK is counterclockwise. To insure that the modulation has the same bandwidth as the legacy IEEE 802.11 DSSS PHY, the chipping rate is kept at 11 Mbit/s while the symbol rate is increased to 1.375 Mbit/s. The spreading rate remains constant, and only the data rate changes. The CCK spectrum is the same as the legacy IEEE 802.11 waveform.

For 5.5 Mbit/s transmission, the scrambled binary bits of the PSDU are grouped into 4-bit nibbles where two of the bits select the spreading function while the remaining two bits QPSK-modulate the symbol as illustrated in Figure 12–10. The spreading sequence then DQPSK-modulates the carrier by driving the I and Q modulators. For 11 Mbit/s operation, the incoming scrambled PSDU binary bits are grouped into 2 bits and 6 bits. The 6 bits are used to select (one of 64) complex vectors of 8 chips in length for the symbol, and the other 2 bits DQPSK-modulate the entire symbol. The transmit waveform is the same, and the chipping rate is maintained at 11 Mbit/s.

DSSS packet binary convolutional coding (PBCC)

PBCC is an optional coding scheme defined in IEEE 802.11b. The coding option uses a 64-state binary convolutional code (BCC), rate R = 1/2 code, and a cover sequence. The PBCC modulator is illustrated in Figure 12–11. The HR/DSSS PMD uses PBCC to transmit the PPDU. To ensure that the PPDU frame is properly decoded at the receiver, the BCC encoder's memory is cleared at the beginning and at the end of a frame. A cover sequence is used to map the QPSK symbols. The cover sequence is initialized with a 16-bit pattern (0011001110001011) to produce a 256-bit cover sequence, which selects the QPSK symbols. BPSK is used for 5.5 Mbit/s and QPSK for 11 Mbit/s. For QPSK, each pair of output bits from the BCC is used to generate one symbol; conversely each pair of bits for BPSK produce 2 symbols. The result is 1 bit per symbol for QPSK and ½ bit for BPSK. Refer to IEEE 802.11b for the equation used for the cover sequence generator.

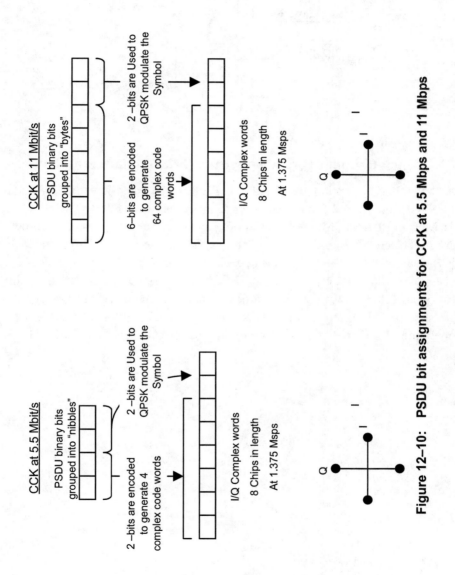

Figure 12–10:   PSDU bit assignments for CCK at 5.5 Mbps and 11 Mbps

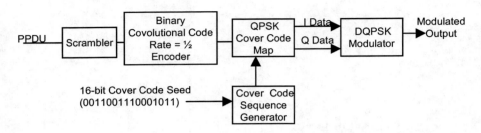

Figure 12–11:   PBCC modulator

Frequency-hopping spread spectrum (FHSS) interoperability

A channel agility option is defined in IEEE 802.11b that allows IEEE 802.11 FHSS 1 Mbit/s and 2 Mbit/s networks to interoperate with HR/DSSS 11 Mbit/s WLANs. Both nonoverlapping and overlapping high-rate channels are supported. The nonoverlapping allows WLANs to operate simultaneously in the same area without interfering each other. In North America, channel 1, channel 6, and channel 11 are specified for nonoverlapping networks, and in Europe (excluding France and Spain) channel 1, channel 7, and channel 13 are specified. Two sets of hopping sequences are defined for worldwide operation. For more details on the hopping patterns, refer to IEEE 802.11b.

Chapter 13 IEEE 802.11j operation in Japan at 4.9 GHz and 5 GHz

In August 2002, the Japanese government adopted a new set of spectrum regulations for operating WLANs in Japan. The new rules extend the frequency band of operation beyond the 5.15–5.25 GHz range to include the both the licensed and unlicensed spectrum from 4.9 GHz to 5.25 GHz. The new regulations opened a new horizon of options of how IEEE 802.11a OFDM devices could be deployed in the licensed and unlicensed spectrum. The new regulations define how WLANs will operate for indoor, fixed outdoor, and mobile deployments. To satisfy these types of deployments, new parameters for the amount of transmit RF power, spurious emission levels, and channel bandwidths were defined. To ensure that legacy IEEE 802.11a OFDM WLANs can interoperate under the new and anticipated regulations, the IEEE 802.11 Working Group formed IEEE 802.11 Task Group j (TGj), chartered to define a set of MAC and PHY specifications to expand the operation of IEEE 802.11a OFDM WLANs into Japan's expanded spectrum and its new applications. The specifications for IEEE 802.11j were defined so that legacy IEEE 802.11a 5 GHz WLANs would require only minimal changes to the PHY hardware and software of the radios. Throughout the development of IEEE 802.11j, the IEEE 802.11 Working Group worked closely with the Multimedia Mobile Access Communications Systems Promotion Council (MMACS-PC); and in September 2004, IEEE 802.11j was ratified by the IEEE Standards Board. This chapter examines some of the key specifications and regulatory domain metrics to be considered for developing devices and WLANs that are compliant with IEEE 802.11j.

EXPANDED COUNTRY INFORMATION ELEMENT

The Country information element is expanded by IEEE 802.11j. This information element follows the same format as described in IEEE 802.11d. The minimum length is 8 octets. The exchange of Country element information is accomplished through a sequence of Beacon and Probe Request frames between the STA and APs prior to joining the network.

Maximum Transmit Power Level field indicates how much transmit RF power a radio is capable of delivering. This was present in IEEE 802.11d.

Two other fields, First Channel Number and Number of Channels, are included in this information element to give the first channel number of the operating frequency band and the number of channels supported. These three fields together are referred to as the *subband triplet*. See Figure 13–1. This sequence is analogous to the transmit power level element request used in IEEE 802.11h.

Element ID	Length	First Channel Number	Number of Channels	Maximum Transmit Power Level
		Subband triplet		

Figure 13–1:   Country information element with subband triplet

Dot11RegulatoryClassesRequired is a MIB attribute that is added by IEEE 802.11j. It includes the reporting of the regulatory and coverage classifications of the STA requesting to join a network. If dot11Regulatory-ClassesRequest is true and if the Regulatory Extension Identifier field value is equal to or greater than a value of 201, then the regulatory class and coverage class are reported following the Regulatory Extension Identifier field. See Figure 13–2. The Regulatory Class field and Coverage Class field have a length of 1 octet each. These three fields together are referred to as the *regulatory triplet*.

Element ID	Length	Regulatory Extension Identifier	Regulatory Class	Coverage Class
		Regulatory triplet		

Figure 13–2:   Country information element with regulatory triplet

The air propagation delay for fixed-access outdoor networks is illustrated in Figure 13–3. The Coverage Class field specifies the aAirPropagationTime value between a STA and AP during data transmission in a network. Coverage Class field was added to accommodate deployment of fixed and mobile AP networks for outdoor environments. For WLANs deployed outdoors, the aAirPropagationTime value may vary from network to network depending upon transmit RF power, range, and environmental conditions. The Coverage Class field and air propagation values shown in Table 13–1 represent an estimated range that is capable for an outdoor network ranging up to ~28 km operating in the 4.900–5.091 GHz licensed band. Coverage class 1 is used for IEEE 802.11a devices. It represents an aAirPropagationTime value of ≤1 μs.

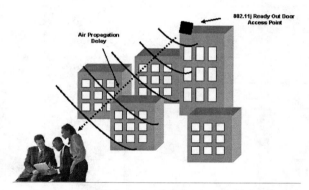

Figure 13–3: Air propagation delay for outdoor fixed access networks

The distance of the air propagation time as shown in Table 13–1 is calculated using the aAirPropagationTime value and the speed of light as is given by the following equation:

$$\text{Distance} = \text{air propagation time} \times \text{speed of light}$$

where

speed of light = ~186,000 mi/s or ~300,000 km/s

The values in Table 13–1 are estimated distance calculations and do not take into account other factors such as path loss, multipath, and signal fading. However, for most WLAN deployments, these factors should be considered in the analysis to guarantee optimal system performance.

Table 13–1:   Coverage Class field values

Coverage Class field value	aAirPropagation time (µs)	Distance (mi)	Distance (km)
1	Less than or equal to 1	0.186	0.300
2	3	0.558	0.900
3	6	1.120	1.800
4	9	1.670	2.700
5	12	2.230	3.600
6	15	2.790	4.500
7	18	3.348	5.400
8	21	3.906	6.300
9	24	4.464	7.200
10	30	5.580	9.000
11	33	6.138	9.900
12	36	6.696	10.800
13	39	7.254	11.700
14	42	7.812	12.600
15	45	8.370	13.500
16	48	8.928	14.400
17	51	9.672	15.600
18	54	10.044	16.200

Table 13–1: Coverage Class field values *(Continued)*

Coverage Class field value	aAirPropagation time (μs)	Distance (mi)	Distance (km)
19	57	10.602	17.100
20	60	11.160	18.000
21	63	11.718	18.900
22	66	12.276	19.800
23	69	12.834	20.700
24	72	13.392	21.600
25	75	13.950	22.500
26	78	14.508	23.400
27	81	15.066	24.300
28	84	15.624	25.200
29	87	16.182	26.100
30	90	16.740	27.000
31	93	17.298	27.900
32–255	—	—	—

MANDATORY AND OPTIONAL MODES OF OPERATION

To insure that IEEE 802.11j networks interoperate with various chipset and equipment providers implementing products compliant to IEEE 802.11j, a set of mandatory specifications is listed in Table 13–2. The mandatory specifications include the following functions: carrier frequency, channel mask, channelization (spacing and width), modulation type, and data rates. Most of the functions are extensions to the IEEE 802.11a specifications. For STA channel synchronization under the MLME-SCAN specification, active scanning of frequency channels by a STA is strictly prohibited and not allowed in the 4.900–5.091 GHz licensed and unlicensed frequency bands.

Table 13–2:   Mandatory modes of operation

Function	Specification	Comments
Carrier frequency	4.9–5.091 GHz, 5.15–5.25 GHz	Operation in Japan
Channel mask	OFDM	Similar to IEEE 802.11a with narrower filter profile for 10 MHz channels
Channel spacing and width	20 MHz	For data rates: 6, 9, 12, 18, 24, 36, 54 Mbit/s
Channel spacing and width	10 MHz	For data rates: 3, 4.5, 6, 9, 12, 18, 24, 27 Mbit/s
Modulation type	OFDM IEEE 802.11a	Mandatory
Data rates	6, 12, 24 Mbit/s	Mandatory for OFDM 20 MHz channel spacing and width
Data rates	3, 6, 12 Mbit/s	Mandatory for OFDM 10 MHz channel spacing and width
STA channel synchronization	MLME-SCAN	Active scanning channel is not allowed in Japan's 4.9–5.25 GHz frequency bands

PLCP HEADER, SIGNAL FIELD, AND RATE SUBFIELD

IEEE 802.11j utilities the same PLCP header specified for IEEE 802.11a radios. Figure 13–4 illustrates the field formats of the PLCP header. The date rate at which the PLCP Preamble and Signal fields are transmitted is based upon the selected channel spacing. For 20 MHz channelization, the PLCP

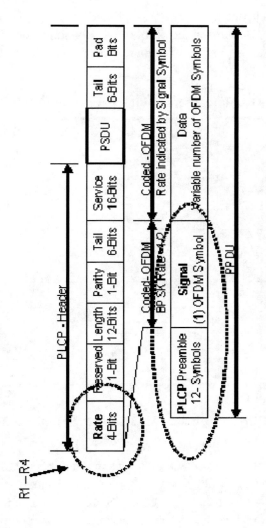

Figure 13–4: PLCP header

Preamble and Signal fields are transmitted using OFDM-BPSK modulation at code rate R = 1/2 at 6 Mbit/s. For 10 MHz channels, the "half-clocked" operation is employed; therefore, the symbol timing is doubled, and the PLCP Preamble and Signal fields are transmitted at 3 Mbit/s. The Rate subfield within the Signal field contains 4 bits (R1–R4) and describes the data rate at which the MPDU will be transmitted. Table 13–3 illustrates the bit assignments as they correspond to the data rate for 10 MHz and 20 MHz channel spacing.

Table 13–3:   Rate subfield bit assignment (R1–R4)

Rate subfield R1–R4	Data rate (Mbit/s) 20 MHz channels	Date rate (Mbit/s) 10 MHz channels
1101	6	3
1111	9	4.5
0101	12	6
0111	18	9
1001	24	12
1011	36	18
0001	48	24
0011	54	27

EXTENDED FREQUENCY BANDS AND TRANSMIT RF POWER LEVELS

Prior to the adoption of Japan's new regulations in August 2002, the 5.15–5.25 GHz frequency band was the only frequency band available for IEEE 802.11a WLANs in Japan. In the United States, Canada, and Europe, IEEE 802.11a WLAN devices are variously permitted to operate from 5.15 GHz to 5.850 GHz or from 5.125 GHz to 5.725 GHz, where the RF transmit power ranges from 40 mW to 800 mW for the different subbands as shown in Table 13–4.

Table 13–4:   Transmit RF power in United States and Europe

Frequency band (GHz)	Tx RF power (maximum)	Europe CEPT (EIRP)
5.125–5.250	40 mW (2.5 mW/MHz)	200 mW
5.250–5.350	200 mW (12.5 mW/MHz)	200 mW
5.470–5.725	800 mW (50 mW/MHz)	1 W
5.725–5.850	800 mW (50 mW/MHz)	N/A

Under the new spectrum rules, two frequency bands were designated for unlicensed and licensed users as shown in Table 13–5. The 5.150–5.250 GHz band is used by legacy IEEE 802.11a WLANs where range is limited to indoor applications and the transmit RF power is constrained to 10 mW/MHz. The other frequency band of operation spans across the 4.9–5.091 GHz band where both licensed and unlicensed users are allowed. If IEEE 802.11j WLANs are deployed outdoors where APs are mounted in fixed locations on building structures or antenna towers, for example, and the transmit RF power is less than 250 mW and less than 50 mW/MHz, then the spectrum is considered to be licensed and offered to a service provider that can provide wireless access to mobile client (i.e., unlicensed) users. Mobile clients in the 4.9–5.091 GHz band are classified as unlicensed users as long as their transmit RF power does not exceed 10 mW/MHz. For more detailed information on the use and restrictions of the licensed spectrum, refer to the regulatory domain documents listed in Table 13–7 (found later in this chapter).

When measuring the maximum transmit RF power allowed in these bands for mobile STAs (i.e., clients) and APs, its important to note that in the United States and Canada, the transmit RF power is measured as conducted RF power at the antenna port; and in Europe and Japan, the measurement is based on equivalent isotropically radiated power (EIRP).

Table 13–5:   Transmit RF power in Japan

Frequency band (GHz)	Regulatory type of deployment	Tx RF Power (EIRP)
4.900–5.091	Fixed access licensed	< 250 mW and < 50 mW/MHz for licensed operation
4.900–5.091	Nomadic mobile access, unlicensed	< 10 mW/MHz
5.150–5.250	Unlicensed	<10 mW/MHz

TRANSMIT MASK AND ADJACENT CHANNEL INTERFERENCE

IEEE 802.11j specifies two spectral masks for OFDM-modulated data transmission in the 4.9–5.0 GHz frequency bands: one for the 20 MHz channel spacing and the other for 10 MHz channel spacing. The shape of the spectral mask shown in Figure 13–5 illustrates the filter response for legacy IEEE 802.11a radios operating on 20 MHz channel spacing. The transmit spectrum is referenced from 0 dBr, which is relative to the maximum spectral density of the transmitted OFDM signal. The bandwidth of the 20 MHz channel does not exceed 18 MHz at 0dBr; and it is offset by 11 MHz at –20 dBr, 20 MHz at –28 dBr, and 30 MHz or higher at –40 dBr from the center frequency. This is the identical mask specified for IEEE 802.11a devices operating in the unlicensed 5 GHz spectrum in the United States, Canada, Europe, and Australia. For IEEE 802.11j, the 20 MHz channel mask remains unchanged. For 10 MHz channels, the spectral shape is narrower and follows the same filter response as the 20 MHz channel. If 10 MHz channels are used, the transmitted spectrum does not exceed 9 MHz of bandwidth at 0 dBr; and it is offset by 5.5 MHz at –20dBr, 10 MHz at –28dBr, and 15 MHz or higher at –40 dBr from the center frequency, as shown in Figure 13–6. The 10 MHz bandwidth is easily obtained because the half-clocked operation is employed in the modem. Half-clocking doubles the OFDM symbol timing and reduces the data rate and required bandwidth by a factor of 2.

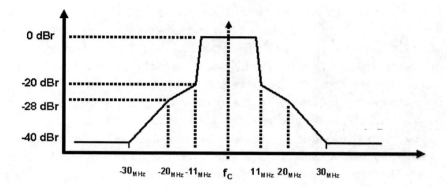

**Figure 13–5:   Transmit mask for 20 MHz channel spacing –
OFDM operation**

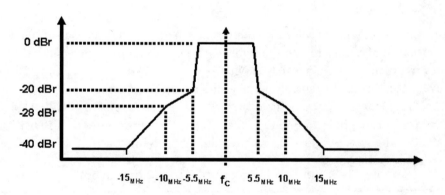

**Figure 13–6:   Transmit mask for 10 MHz channel spacing –
OFDM operation**

The 4.9–5.0 GHz frequency band in Japan is defined for fixed-access
licensed, mobile access unlicensed, and general WLAN unlicensed use;
therefore, spectrum interference from the adjacent channel and co-channel
needs to be carefully managed in the radio. For IEEE 802.11j devices

operating in Japan on the channels specified across the 4.9–5.0 GHz frequency band, the average transmit power emissions in an adjacent channel need to be –3 dBr and –18 dBr for alternate adjacent channels having the same channel bandwidth (i.e., 10 MHz or 20 MHz). This measurement is made relative to EIRP in the 100 kHz bandwidth at the channel edges and the band edges. However, for operation in Japan at the 5.15–5.25 GHz frequency band, the average power emissions should be less than the average on-channel power by 25 dB and 40 dB, respectively. These measurements are critical for achieving regulatory domain certification in Japan. Transmit mask and adjacent channel measurements are made relative to EIRP for operation in Japan. These measurements are typically made with a spectrum analyzer set at a 100 kHz resolution bandwidth with a 30 kHz video bandwidth along with a RF power meter. For more detailed information on how to test conformance and to measure spectral emissions for Japan's regulatory domain, refer to MPHPT Equipment Ordinance for Regulating Radio Equipment articles.

SPURIOUS EMISSIONS

In addition to adjacent channel interference, both in-band and out-of-band spurious emissions are important measurements for compliance under the new regulations in Japan. Just as with legacy IEEE 802.11a STA and APs, every IEEE 802.11j transmitting over the RF medium must comply with the local regulations. For compliance in the United States, refer to 47CFR 15.407; and for Europe, refer to ETSI EN 301-389-1. For Japan, refer to MPHPT EO 49.20 and 49.21, Section 10, as shown in Table 13–6.

Table 13–6:   In-band and out-of-band spurious emissions

Regulatory domain	Frequency band (GHz)	Leakage power (out-of-band)	Frequencies (out-of-band) (MHz)
Japan E. O. Article 49.21 Section 10	4.9	< 15 μW/MHz	$4880 \leq f < 4900$ $5000 < f \leq 5020$

Table 13–6:   In-band and out-of-band spurious emissions *(Continued)*

Regulatory domain	Frequency band (GHz)	Leakage power (out-of-band)	Frequencies (out-of-band) (MHz)
Japan E. O. Article 49.21 Section 10	5.0	< 15 µW/MHz	$5000 \leq f < 5020$ $5100 < f \leq 5120$
Japan E. O. Article 49.21 Section 10	5.0	< 0.5 W/MHz	$5020 \leq f < 5030$ $5091 < f \leq 5100$

REGULATORY DOMAIN REFERENCES

All IEEE 802.11j OFDM STA and AP devices must comply with Japan's regulatory body and achieve certification for the radio from MPHPT before operating in the 4.9-5.25 GHz frequency bands. This chapter covers a number of the key specifications of the PHY specifications relative to implementing IEEE 802.11j STA and APs, but does not cover the details of the new spectrum rules. It is recommended that documents listed in Table 13–7 (in particular, MPHPT Articles 7, 49.30, and 49.21) be considered early on in the product development cycle of designing a IEEE 802.11j product or system.

Table 13–7:   Regulatory bodies for 4.9–5 GHz spectrum

Geographic region	Government body	Regulatory documents	Approval authority
Japan	Ministry of Public Management, Home Affairs, Posts and Telecommunications (MPHPT)	MPHPT – Equipment Ordinance For Regulating Radio Equipment (E.O.) Articles 7, 49.20, 49.21	MPHPT
United States	FCC	CFR 47 [B6], Part 15, Sections 15.205, 15.209, and 15.247 and Subpart E, Sections 15.401 – 15.407	FCC

Table 13–7: Regulatory bodies for 4.9–5 GHz spectrum *(Continued)*

Geographic region	Government body	Regulatory documents	Approval authority
Canada	Industry Canada	ICES-003 subpart B, BGETS-7 subpart B, RSS-210 subpart C, RSS-213 subpart D	Industry Canada
Europe	ETSI	ERC DEC (99)23, ETSI EN301-893	CEPT

NUMBER OF FREQUENCY CHANNELS AND DATA RATES

IEEE 802.11j follows the same channel numbering scheme specified for IEEE 802.11a and extends it to allow for a channel starting frequency to be communicated separately. The original IEEE 802.11a channel numbering scheme accounted only for frequencies from 5 GHz to 6 GHz, and the new numbering scheme allows for any arbitrary frequency. Note that the channel spacing is different for nonoverlapping adjacent channel center frequencies. The channel center frequency is calculated as follows:

$$\text{Channel center frequency} = \text{Starting frequency} + (N_{channels} \times 5 \text{ MHz})$$

$$\text{Starting frequency, i.e., 4000, 5000}$$

where

$$N_{channels} = 0, 1, 2, \ldots.200$$

IEEE 802.11j supports the data rates for IEEE 802.11a operating on 20 MHz channel spacing, and by using the half-clocked operation, data rates are cut in half when operating on 10 MHz channel spacing. By using half-clocked operations, the symbol timing is doubled and the data rates transmitted and received at half the IEEE 802.11a rate. The mandatory data rates are illustrated in Table 13–8 for the respective channel spacing.

Table 13–8:   Mandatory data rates

Channel spacing (MHz)	Data rates (Mbit/s)	Mandatory data rates (Mbit/s)	Comments
20	6, 9, 12, 18, 24, 36, 48, 54	6, 12, 24	Support for legacy IEEE 802.11a rates
10	3, 4.5, 6, 9, 12, 18, 24, 27	3, 6, 12	

RECEIVER SENSITIVITY, CCA, AND SLOT TIME

The minimum receiver sensitivity for IEEE 802.11j devices is specified in Table 13–9 for operation on 10 MHz and 20 MHz channels. For 20 MHz channel spacing, the minimum receiver sensitivity values from IEEE 802.11a are specified. There is a ~16 dB difference between the adjacent channel rejection (ACR) and the alternate adjacent channel rejection and a 3 dB difference in the minimum receiver sensitivity between 10 MHz and 20 MHz channels. The minimum receiver sensitivity in a 20 MHz channel is –82 dB at 6 Mbit/s and –65 dB at 54 Mbit/s. For a 10 MHz channel, the minimum receiver sensitivity improves by 3 dB. The packet error rate (PER) is measured at the minimum receiver sensitivity values. The PER at minimum receiver sensitivity is 10% at a packet length of 1000 bytes. The minimum levels are measured at the antenna port on the radio. A noise figure (NF) loss of 10 dB and implementation losses of 5 dB are assumed.

The slot time for operating is 9 μs in a 20 MHz channel and 13 μs in 10 MHz channels, as shown in Table 13–10. However, if dot11RegulatoryClasses-Required is active, then the slot time will increase by 3 × the desired coverage class. The times for CCA, preamble length, and PLCP header length increase by 2x for 10 MHz channel spacing. When designing an IEEE 802.11j WLAN for operation in Japan, special attention should be given when calculating the throughput of a network and client loading per AP for operation in the 4.9–5.250 GHz band. As shown in Table 13–10, the protocol timing parameter values for media transmission are different for 10 MHz and 20 MHz channels.

Table 13–9:   Receiver sensitivity for 10 MHz and 20 MHz channel spacing

Modulation	Coding rate	ACR (dB)	Alternate ACR (dB)	Receiver sensitivity (minimum) 20 MHz channels	Receiver sensitivity (minimum) 10 MHz channels	Data rate (Mbit/s) 10 MHz channels	Data rate (Mbit/s) 20 MHz channels
BPSK	1/2	16	32	−82	−85	3	6
BPSK	3/4	15	31	−81	−84	4.5	9
QPSK	1/2	13	29	−79	−82	6	12
QPSK	3/4	11	27	−77	−80	9	18
16-QAM	1/2	8	24	−74	−77	12	24
16-QAM	3/4	4	20	−70	−73	18	36
64-QAM	2/3	0	16	−66	−69	24	48
64-QAM	3/4	1	15	−65	−68	27	54

Table 13–10:   Protocol timing parameters

Parameter	20 MHz channelspacing (µs)	10 MHz channel spacing (µs)
Slot time	9	13
SIFS time	16	32
CCA time	< 4	< 8
Preamble length	20	40
PLCP header length	4	8

TRANSMITTER ERROR VECTOR MAGNITUDE (EVM)

Transmitter EVM is a root-mean-square (rms) measure of the fidelity or relative linearity of the transmitted RF signal over the RF media compare to what is theoretically possible for the modulation employed. In IEEE 802.11j, EVM is referred to as transmit[ter] constellation error. IEEE 802.11j uses the same values for EVM, i.e., the relative constellation error values, as specified for IEEE 802.11a. See Table 13–11.

Table 13–11:   EVM vs data rate

Date rate in 10 MHz channel (Mbit/s)	Data rate in 20 MHz channel (Mbit/s)	EVM relative constellation error (dB)
3	6	−5
4.5	9	−8
6	12	−10
9	18	−13
12	24	−16
18	36	−19

Table 13–11:   EVM vs data rate *(Continued)*

Date rate in 10 MHz channel (Mbit/s)	Data rate in 20 MHz channel (Mbit/s)	EVM relative constellation error (dB)
24	48	–22
27	54	–25

Chapter 14 IEEE 802.11g higher data rates in 2.4 GHz frequency band

Since the ratification of IEEE 802.11b and IEEE 802.11a in 1999, the demand for IEEE 802.11b Wi-Fi networking has grown to be a huge success for home computing and enterprise market. Much of the demand for IEEE 802.11b wireless connectivity has triggered the need for higher data rates to support applications, such as streaming video, and the user capacity and throughput that is key in enterprise and hotspot deployments.

In June of 2003, IEEE 802.11g was ratified by the IEEE. IEEE 802.11g is the newest PHY extension that enables speeds ranging from 1 Mbit/s to 54 Mbit/s operating in the 2.4 GHz frequency band to use the existing MAC. IEEE 802.11g specifies complementary code keying (CCK) from IEEE 802.11b and orthogonal frequency division multiplexing (OFDM) from IEEE 802.11a as mandatory modulations for the PHY. New mechanisms where put in place in the MAC protocol to maintain backwards compatibility and interoperability with legacy IEEE 802.11b WLANs. The IEEE 802.11 Working Group decided to use OFDM in the frequency band in 2001 after the FCC in the United States approved the use of OFDM signaling in 2.4 GHz unlicensed frequency band. Up until May 2001, only IEEE 802.11b Wi-Fi devices using Barker word spreading, CCK, or PBCC modulation were allowed in this band. For more information on the details of IEEE 802.11b CCK and IEEE 802.11a OFDM, refer to Chapter 12.

NETWORK DEPLOYMENT AND USER SCENARIO

IEEE 802.11g is designed to support three types of WLAN scenarios:

- Legacy IEEE 802.11b CCK,
- Mixed-mode IEEE 802.11b CCK and IEEE 802.11g OFDM, and
- Green field (i.e., new) installations with IEEE 802.11g OFDM.

See Figure 14–1. Legacy Wi-Fi devices are limited to receiving and detecting CCK signals and not OFDM signals. Initial deployments of IEEE 802.11g wireless systems will operate as mix-mode networks, where legacy IEEE 802.11b CCK STAs and IEEE 802.11g CCK and OFDM STAs coexist and interoperate with IEEE 802.11g APs and fall back to CCK at data rates < 11 Mbit/s. However, for new green field deployments where all STAs are IEEE 802.11g devices capable of receiving either CCK or OFDM transmissions, data rates up to 54 Mbit/s are obtainable, and comparable throughputs to IEEE 802.11a devices operating in the 5 GHz band are achievable.

MANDATORY AND OPTIONAL MODES OF OPERATION

IEEE 802.11b CCK and IEEE 802.11a OFMD signaling are mandatory for the IEEE 802.11g PHY. See Table 14–1. Each STA must support three preamble and PLCP header PPDU formats. The mandatory data rates for transmitting and receiving data payloads are 1 and 2 Mbit/s for DSSS; 5.5 and 11 Mbit/s for CCK; and 6, 12, and 24 Mbit/s for OFDM. For DSSS and CCK, it is mandatory for all STAs and APs to lock the transmit center frequency and symbol clock frequency to the same reference oscillator. The optional data rates supported for OFDM are 9, 18, 36, 48, and 54 Mbit/s. Although 54 Mbit/s is an optional data rate for compliance to IEEE 802.11g, it is mandatory for IEEE 802.11g devices obtaining Wi-Fi certification. For most applications, such as the transmission of MPEG-2 video using wireless standard definition television (SDTV), a data rate of 24 Mbit/s is sufficient. Other mandatory features for CCK include

- Support of short preamble,
- Maximum RF input signal limited to –20 dBm,
- Transmit center frequency and symbol clock frequency locked to the same reference oscillator.

Lastly, the CCA mechanism must detect all mandatory sync symbols.

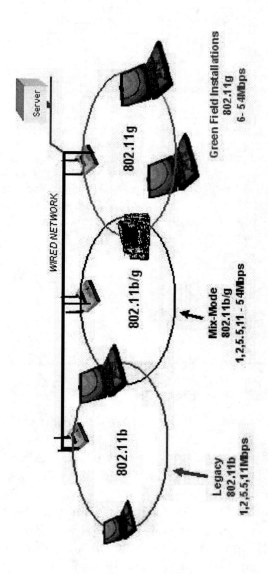

Figure 14-1: IEEE 802.11g system deployment

Table 14–1:   Mandatory modes of operation

Function	Specification	Notes
Carrier frequency	2.400–2.4835 GHz	Mandatory – worldwide
Channel mask	CCK, OFDM	Same as IEEE 802.11a and IEEE 802.11b
Channel spacing	25 MHz channel spacing	Same as IEEE 802.11b
Modulation type	DSSS, CCK, and OFDM	Mandatory (CCK short and long preamble)
Data rates	1, 2, 5.5, 6, 11, 12, and 24 Mbit/s	Mandatory (1 and 2 Mbit/s for DSSS, 5.5 and 11 Mbit/s for CCK, 6, 12, and 24 Mbit/s for OFDM)
Data rates	9, 18, 36, 48, and 54 Mbit/s	Optional
CCA detection	CCK and OFDM preamble sync symbol	Mandatory
Transmitter clock lock	DSSS-CCK	Mandatory (center frequency and symbol clock)

OPTIONAL MODES OF OPERATION

IEEE 802.11g specifies CCK-OFDM and PBCC as optional modes of operation. CCK-OFDM and PBCC are allowed by the IEEE 802.11 standard, but are not required to implement IEEE 802.11g STAs or APs. CCK-OFDM is referred to as a hybrid waveform that combines CCK (i.e., single carrier) and OFDM (i.e., multicarrier) modulations for transmission of IEEE 802.11g packets. The preamble and PLCP header are transmitted using CCK, and the PSDU payload is transmitted using OFDM. The short and long preambles used in IEEE 802.11b are specified for CCK-OFDM transmission. See Figure 14–2 and Figure 14–3. This feature allows legacy IEEE 802.11b Wi-Fi devices to receive and detect the preamble of the CCK portion of the CCK-

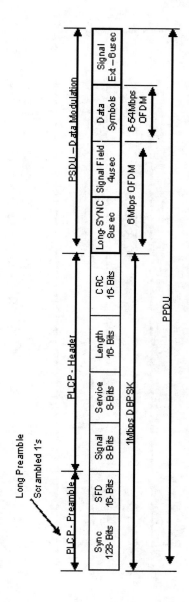

Figure 14–2:   CCK-OFDM hybrid: long preamble

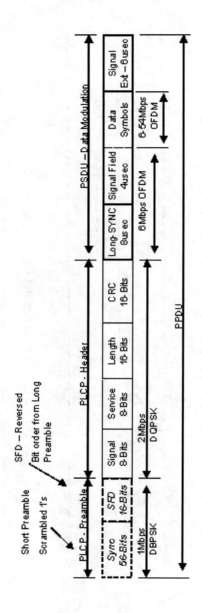

Figure 14–3: CCK-OFDM hybrid: short preamble

OFDM transmission and defer the medium to prevent the possibility of collisions. Therefore, coexistence between CCK-OFDM and legacy IEEE 802.11b Wi-Fi devices on the channel is not a problem.

During transmission of a long preamble, the PLCP is transmitted at 1 Mbit/s using DBPSK modulation for the PPDU, and the PSDU payload transmits the long OFDM symbols and Signal field at 6 Mbit/s and the data symbols at the payload rate specified in the Signal field (i.e., 6 Mbit/s to 54 Mbit/s). During transmission of the short preamble in the PPDU, the 56-bit Sync field and the Start of Frame Delimiter (SFD) field are transmitted at 1 Mbit/s using DBPSK modulation, and the PSDU is transmitted at the OFDM rates (i.e., 6 Mbit/s to 54 Mbit/s).

PBCC is the second optional mode of operation. PBCC is also an option in the legacy IEEE 802.11b PHY specification. Like CCK-OFDM, PBCC is considered a hybrid waveform. PBCC employs a complex signal constellation, 8-PSK, structured using a convolution coding technique that builds on the transmission of CCK preambles. PBCC transmits the preamble and PLCP header using CCK, and the payload is transmitted using PBCC. PBCC supports the basic rates of 1, 2, 5.5, and 11 Mbit/s and employs a code structure that extends the data rates to 22 Mbit/s and 33 Mbit/s. See more details on PBCC see Chapter 12.

PPDU FORMATS

All IEEE 802.11g STAs must support three types of preamble and PLCP headers as follows:

a) Long preamble and PLCP header (See Figure 14–4 for IEEE 802.11g Service field)

b) Short preamble and PLCP header

c) OFDM preamble and PLCP header

The Service field bit positions are shown in Figure 14–4. Data bits b0, b1, and b4 are reserved bits and always set to logic 0. The Service field uses 3 bits of the reserved 8 bits for IEEE 802.11b. Data bit b2 determines whether the transmit frequency and symbol frequency clocks use the same local oscillator. Setting this bit to logic 1 is mandatory for all IEEE 802.11g STAs and APs.

b0	b1	b2	b3	b4	b5	b6	b7
Reserved	Reserved	Locked Clock 1 = locked	Modulation Selection 0 = CCK 1 = ERP-PBCC	Reserved	Length Extension ERP-PBCC 22-33	Length Extension ERP-PBCC 22-33	Length Extension

Figure 14–4:   Service field based on IEEE 802.11g

Data bit b3 indicates whether CCK or PPBC is used. Data bit b7 is a bit extension used in conjunction with the Length field to calculate the duration of the PSDU, in microseconds, for CCK if data bits b3, b5, and b6 are set to logic 0. If data bit b3 is set to logic 1, then data bits b5, b6, and b7 are used to resolve data field length resolution, in microseconds, for ERP-PBCC 22 Mbit/s and ERP-PBCC 33 Mbit/s optional modes. To select ERP-PBCC 22 Mbit/s, the Service field is set to DCh; for ERP-PBCC 33 Mbit/s, 21h. For all data rates for DSSS-OFDM, the Service field is set to 1Eh. The data bits in the Service field are transmitted LSB first.

OPERATING CHANNELS

The spectral mask for IEEE 802.11g in the 2.4 GHz frequency utilizes the masks defined for IEEE 802.11b CCK and IEEE 802.11a OFDM. For details on the filter requirements and shape of the masks, see Chapter 12 and Chapter 13. The transmit masks are designed to allow three noninterfering channels spaced 25 MHz apart over the 2.4 GHz frequency band. This feature allows CCK and OFMD STAs and APs to coexist spectrally with legacy IEEE 802.11b Wi-Fi networks and provides a seamless upgrade path to employ IEEE 802.11g. Figure 14–5 illustrates the channel mask and spacing using channel 1, channel 6, and channel 11 for an IEEE 802.11g system using OFDM. The 3 dB channel bandwidth of the OFDM channel is 20 MHz. Therefore, using the legacy channelization plan, with 25 MHz channel spacing, adjacent channel interference is not an issue. However, when operating at the lower and upper band edges of the 2.4 GHz frequency band, additional filtering or backoff in the RF transmit power may be necessary to conform with the local regulatory requirements. Figure 14–6 shows channel 6 operating as an IEEE 802.11b CCK network. In this example, channel 1 and channel 11 are operating as OFDM networks and illustrate the capability of mixed-mode operation.

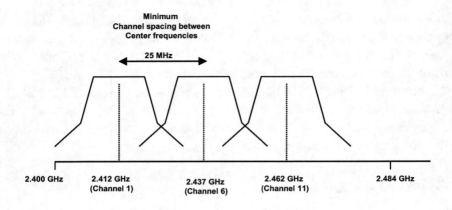

Figure 14–5:   Channel mask spacing for OFDM operation

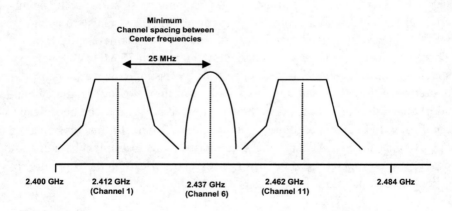

Figure 14–6:   Channel mask spacing using legacy IEEE 802.11b CCK system on channel 6

OPERATION OF IEEE 802.11g CSMA/CA AND CCA

Just as with IEEE 802.11a and IEEE 802.11b, IEEE 802.11g uses the "listen before talk" (LBT) mechanism built into the MAC protocol to transmit packets over the medium. This mechanism is well-known as *CSMA/CA* where a STA must listen and sniff the medium for a clear channel before transmitting packets. This procedure is performed by using the clear channel assessment (CCA) mechanism. For IEEE 802.11g, all STAs must support the three preamble and PLCP header mandatory formats as previously described. This support ensures that all implementations are capable of receiving and detecting DSSS, CCK and OFDM signals and the CCA mechanism must be capable of detecting a medium busy for each PLCP and PPDU type.

CCA detection is based on using the combination of energy detect (ED) and carrier sense (CS), and CS in the receiver must be capable of detecting all Barker and OFDM synchronization symbols in the preambles. Without this capability, both packet collisions and lack of coexistence between CCK and OFDM STAs would occur and potentially severally impair the medium. In a given BSS, which consists of an AP and all associated STAs, the STAs must listen for transmissions to determine whether the channel is idle. If the channel is idle (i.e., where no minimal energy or carrier is detected), then an internal backoff timer begins to count down. Once the timer expires, the STA begins transmission. To insure that the possibility of collisions from other STAs in the BSS is minimized, other mechanisms are employed in the backoff timer to randomize the duration of the timer. However, on the other hand, if the receiver's CCA mechanism is triggered by ED, which detects the minimum signal power threshold equal to or greater than -76 dBm, and if the minimum ED is met and the receiver demodulates a DSSS or OFDM preamble, CCA is reported as medium busy. Then, transmission by the STA requesting to transmit information waits for a clear medium using the internal backoff timer.

Under normal operation, all STAs sharing the medium can hear and detect each other. However, in the case of the hidden node problem, CCK devices cannot hear OFDM transmissions. To overcome this problem, the request to send and clear to send (RTS/CTS) mechanism built in the IEEE 802.11 standard is used for mixed-mode deployments where legacy CCK and OFMD exist in a BSS. When the CTS/RTS mechanism is employed, most STAs in a BSS will hear and detect RTS, and all STAs will hear the CTS from the AP. During a RTS/CTS exchange, each STA receives information on how long the CCK or OFDM packet, including the acknowledge (ACK) transmission, will be. The network allocation vector (NAV), which is an internal timer in each STA, is set to have the same duration as the OFMD packet exchange. The NAV timer and the internal backoff timer operate together as part of the CCA virtual carrier sense mechanism. If the NAV and backoff timer expire and if no energy or CS is detected, the channel is claimed to be idle, and the STA begins to contend for the medium access. This operation allows the coexistence between CCK and OFDM radios in a BSS. Figure 14–7 shows the packet exchange for CTS/RTS with CCK and OFDM transmission with ACK exchanges. It is also noted that all STAs and the AP in a IEEE 802.11g BSS must be capable of gear-shifting back in data rate and operate as a legacy IEEE 802.11b device.

KEY SYSTEM SPECIFICATIONS

Table 14–2 lists a number of key specifications required for CCK and OFDM operation for IEEE 802.11g. The specifications were derived from the IEEE 802.11b CCK and IEEE 802.11a OFDM PHY and MAC protocol specifications.

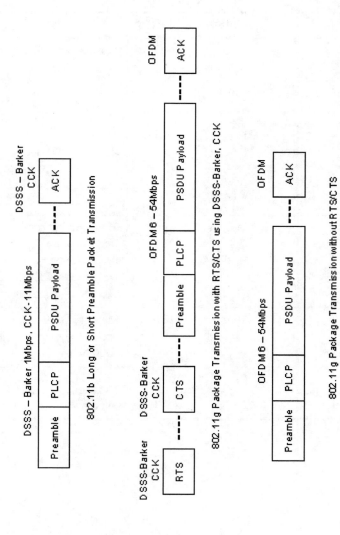

Figure 14–7: IEEE 802.11g CCK-OFDM packet transmission

Table 14–2: Key system parameters

Function	CCK	OFDM	Notes
Slot time	20 μs	9 μs	
CCA	15 μs	4 μs	Must detect all OFMD sync symbols
Probability of detection-CCK	> 99%	> 90%	
Symbol frequency clock	± 25 ppm	± 25 ppm	Transmit and carrier frequency are derived from the same clock for DSSS and CCK (Locked oscillators)
Transmit center frequency accuracy	± 25 ppm	± 25 ppm	Transmit and carrier frequency are derived from the same clock for DSSS and CCK (Locked oscillators)
PER	10%	10%	Measured with PSDU packet length of 1000 bytes
SIFS	10 μs	10 μs	
Receiver input level	–20 dBm	–20 dBm	Maximum

Chapter 15 IEEE 802.11n higher data rates beyond 54 Mbit/s

At the time this book was written, the IEEE 802.11 Working Group was underway developing IEEE 802.11n, a new high-throughput extension to IEEE 802.11 standard. IEEE 802.11n is expected to deliver data rates in excess of 100 Mbit/s in the 2.4 GHz and 5 GHz frequency bands and expected to support backwards compatibility to legacy IEEE 802.11a, IEEE 802.11b, and IEEE 802.11g PHYs and the MAC. The focus of the working group is to define a modulation scheme and a set of specifications for the PHY required to deliver over 100 Mbit/s of data throughput. Although the focus is on the PHY, modifications and new specifications may also be considered for the MAC to support such data rates. IEEE 802.11n is expected to be ratified in 2006.

As IEEE 802.11 Wi-Fi deployments continue to grow in the future, the need for higher data rates beyond 54 Mbit/s in excess of 100 Mbit/s is expected to be driven by the demand of several factors:

a) Portable multimedia devices;

b) Standard definition television (SDTV), high-definition television (HDTV), and consumer electronics in the home;

c) Dense hotspot or enterprise deployment of STAs (i.e., clients) per AP;

d) User network capacity at a premium for mixed data, including Voice Over Internet Protocol (VoIP) and streaming video.

In addition to these demands, it is expected that it will be necessary to maintain the performance and seamless connectivity to legacy IEEE 802.11a, IEEE 802.11b, IEEE 802.11g, and dual-band IEEE 802.11 WiFi devices.

To deliver such data rates, multipath, intersymbol interference (ISI), range, and coexistence with legacy Wi-Fi devices and other users of the 2.4 GHz and 5 GHz frequency bands must be carefully considered. There are three PHY (i.e., radio) technologies under consideration for IEEE 802.11n: channel bonding, higher order modulations, and the multiple input multiple output (MIMO) antenna system.

CHANNEL BONDING

Channel bonding is a radio technique where two or more adjacent frequency channels are combined together to achieve higher data rates. This technique is accomplished by widening the channel bandwidth from 20 MHz to 40 MHz to where the frequency spectrum occupies two or more frequency channels in the 2.4 GHz or 5 GHz frequency band as shown in Figure 15–1. For example, if we consider the 2.4 GHz frequency band, by combining two frequency channels and virtually doubling the channel bandwidth, the OFDM data rate of 54 Mbit/s is doubled to 108 Mbit/s. Data rates higher than 108 Mbit/s can be achieved by using advanced data coding algorithms. On the surface, this technique appears to require minimal changes in the radio implementation. However, there is penalty for using this technique. By doubling the bandwidth, the number of overlapping adjacent channels available is reduced and could potentially cause adjacent channel interference in existing infrastructure deployments. Careful deployment and channel planning must be considered for this radio architecture. This technique is viewed by many as inefficient use of frequency spectrum, and in some regulatory domains in the world the use of technique is not allowed.

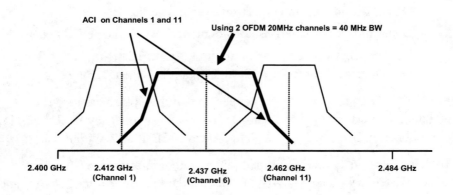

Figure 15–1: Channel bonding in 2.4 GHz frequency band

HIGHER ORDER MODULATION

IEEE 802.11a and IEEE 802.11g use 64-QAM OFDM as the modulation for achieving a data rate throughput of 54 Mbit/s. Higher order modulations, such as 128-QAM or 256-QAM, combined with advanced digital trellis coding techniques could be used to extend the data rates beyond 54 Mbit/s to very high data rates. It is theoretically possible to use the existing IEEE 802.11a and IEEE 802.11g channel bandwidths and channel scheme and stay within the bounds of Shannon's limit to achieve data rates higher than 100 Mbit/s. However, the challenge lies in the specifications of the linearity requirements for the radio's transmitter path, including the power amplifier. For higher order modulations, there is little margin in the transition between symbols in the constellation and minimizing the phase noise in the synthesized LO required to maintain an acceptable error vector magnitude (EVM) and signal-to-noise ratio (SNR) to achieve the range of legacy IEEE 802.11a and IEEE 802.11g Wi-Fi STAs becomes questionable. Furthermore, the channel equalizer in the receiver becomes more complex as the order of the modulation increases to combat and mitigate against signal fading, ISI, and multipath in the channel.

MULTIPLE INPUT MULTIPLE OUTPUT (MIMO)

In a MIMO WLAN, multiple antennas are used at the receiver and at the transmitter as shown in Figure 15–2. Each receive antenna receives the direct path plus the delayed paths and reflected paths from each transmit antenna. MIMO has been around for at least two decades and has become very attractive to IEEE 802.11 because many of the IEEE 802.11a and IEEE 802.11g Wi-Fi devices on the market today use multiple receive antennas to implement some form of antenna spatial diversity in the radio. However, to achieve the maximum channel capacity gains and highest data rates theoretically possible beyond 54 Mbit/s, both transmit and receiver diversity must be considered. There are a number of benefits in using MIMO systems for WLAN:

a) The resistance to fading environments is improved using transmit diversity.

b) There is a lower probability of interference.

c) There is higher channel throughput capacity.

d) MIMO enables the receiver to recover data using a number of multipath signals.

e) MIMO allows for distribution of RF transmit power over multiple antennas.

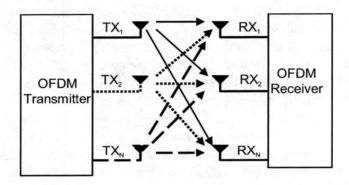

Figure 15–2:   MIMO system with TX$_N$ transmit and RX$_N$ receive antennas

In the receiver, rake or other complex channel equalizers could be employed to recover symbols and timing from the multiple delayed and multipath signals, just as we have seen in IEEE 802.11b systems. Other signal processing techniques, such as maximal ratio combining, could be used in the digital baseband processing to recover data and clock information from the multiple receiver paths. By using MIMO for higher data rates, it is certainly theoretically possible to achieve the same range at 100 Mbit/s as with IEEE 802.11a and IEEE 802.11g 54 Mbit/s systems. However, there are cost and power penalties in achieving such performance with today's semiconductor process technology. The complexity of the radio using MIMO for STAs (i.e., client devices) increases significantly over non-MIMO architectures. At a minimum, there are two transmitters and two receivers that yield an increase in cost and power consumption. Then, depending upon the configuration of the architecture and desired performance in terms of data rate, PER, and range, the number transmitters or receivers could increase. Such an increase would yield yet another level of complexity. However, if we look at the future of higher rate IEEE 802.11 WLAN devices and apply Moore's Law into the economics of semiconductor process technology, we could expect to see the cost of WLAN STAs employing MIMO approach trigger points that will enable the adoption of the technology for the mass market — hopefully by the time IEEE 802.11n is ratified.

Chapter 16 System design considerations for IEEE 802.11 WLANs

The IEEE 802.11 standard provides a number of PHY options in terms of data rates, modulation types, and spreading spectrum techniques. Selecting the right PHY and MAC technologies for your application requires careful planning and detailed systems analysis for developing the optimal WLAN implementation. It is impossible to include every possible system consideration in this handbook. However, we have focused on a few key issues we believe are important for consideration when implementing a IEEE 802.11 interoperable WLAN. The issues covered in this chapter are some of issues on which the IEEE 802.11 Working Group focused during the development of the standard.

THE MEDIUM

The difference between "wired" and RF WLANs is the radio communications link. While the radio communications link provides the freedom to move without constraints of wires, the wired media has the luxury of a controlled propagation media. Wireless RF media are very difficult to control because the dynamics of the propagated signals over the media are constantly changing. This situation is the case for IEEE 802.11 WLANs because the 2.4 GHz bands are shared by unlicensed users. Radio system designers need to have a thorough understanding of the RF medium to properly design 2.4 GHz and 5 GHz IEEE 802.11 WLANs, especially for networks operating at data rates greater than 2 Mbit/s. The RF communication media for home, enterprise, and manufacturing environments are very different, and no two environments are the same. Multipath and path loss are issues to consider when designing an IEEE 802.11 WLAN.

Multipath

Multipath is one of the performance concerns for indoor IEEE 802.11 WLANs. Multipath occurs when the direct path of the transmitted signal is combined with paths of the reflected signal paths and results in a corrupted signal at the receiver, as shown in Figure 16–1. The delay of the reflected signals is measured in nanoseconds (nsec) and is commonly known as *delay spread*. Delay spread is the parameter used to signify multipath. The amount of delay spread varies for indoor home, office, and manufacturing environments, as shown in Table 16–1. Surfaces of furniture, elevator shafts, walls, factory machinery, and metal-constructed buildings all contribute to the amount of delay spread in a given environment.

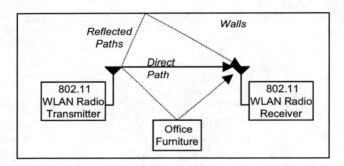

Figure 16–1: How multipath is generated

Table 16–1: Typical multipath delay spread for indoor environments

Environment	Delay spread
Home	< 50 nsec
Office	~ 100 nsec
Manufacturing floor	200–300 nsec

The channel impulse response is a way to illustrate the amount of multipath dispersion. For example, the amount of delay spread in an office environment is approximately 100 nsec, as shown in Figure 16–2. Typically, energy reflected off the surface of walls causes the impulse response to have energy on the leading edge before the peak. The leading energy is called the *precursor energy*. The amount of precursor energy differs from one environment to the next. The processing required to correct the precursor energy is more complex than required for the trailing edge energy. The symbol period on the *x*-axis of the graph in Figure 16–2 is equal to the length of the 8-chip CCK code word. The 11-chip Barker code is only 3 chips longer.

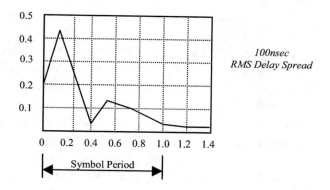

Figure 16–2: Channel impulse response multipath (delay spread) for office environment

Rake processing and equalization are two methods used to process and resolve delay spread. A rake receiver is well-known architecture used to remove delay spreads on the order of 100 nsec. The rake is structured as a bank of correlators (i.e., fingers) with weighed delays and a combiner. Equalization is an alternative used to correct delay spreads greater that 100 nsec. Multipath causes the signals from the previous symbol to interfere with the signals of the next. This is known as intersymbol interference (ISI). As with ISI, interchip interference (ICI) results when the signals of the previous chip interfere with the signals of the next chip. ISI and ICI are issues for higher data rate systems because the symbol and chip periods are shorter.

This situation is the case for IEEE 802.11a and IEEE 802.11b. Equalization corrects for ISI and ICI. An equalizer is a multitapped delay line, which takes the delayed and attenuated signal and subtracts it from the actual received signal. However, for environments where delay spreads are greater than 200 nsec, more complex signal processing is necessary. Rake processing combined with ISI and ICI equalization is commonly implemented to resolve multipath dispersions of this magnitude.

Multipath channel model

In an environment where performance measurements of the same radio are used, in the same location, the results may not agree. The changing positions of people in the room and slight changes in the environment can produce significant changes in the signal power at the radio receiver. A consistent channel model is required to allow comparison of different WLANs and to provide consistent results. In doing so, the IEEE 802.11 Working Group adopted the channel model below as the baseline for predicting multipath for modulations used in IEEE 802.11a (5 GHz) and IEEE 802.11b (2.4 GHz). This model is ideal for software simulations predicting performance results of a given implementation. The channel impulse response illustrated in Figure 16–3 is composed of complex samples with random, uniformly distributed phase and Rayleigh-distributed magnitude with average power decaying exponentially.

The mathematical model for the channel is as follows:

$$h_k = N(0, \tfrac{1}{2}\sigma^2_k) + jN(0, \tfrac{1}{2}\sigma^2_k)$$

$$\sigma^2_k = \sigma^2_0 e^{-kT_s/T_{RMS}}$$

$$\sigma^2_0 = 1 - e^{-T_s/T_{RMS}}$$

where

$N(0, \tfrac{1}{2}\sigma^2_k)$ is a zero mean Gaussian random variable with variance $\tfrac{1}{2}\sigma^2_k$ produced by generating an $N(0,1)$ and multiplying it by $\sigma_k / \sqrt{2}$), and $\sigma^2_0 = 1 - e^{-T_s/T_{RMS}}$ is chosen so that the condition $\sum \sigma^2_k = 1$ is satisfied to ensure same average received power.

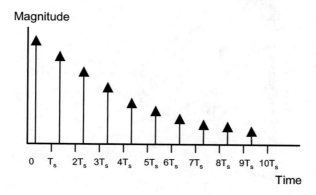

**Figure 16–3:   Channel impulse response
for IEEE 802.11a and IEEE 802.11b**

Let T_s be the sampling period and T_{RMS} be the delay spread of the channel. The performance assessment shall be no longer than the smaller of 1/(signal bandwidth) or $T_{RMS}/2$. The number of samples to be taken in the impulse response should ensure sufficient decay of the impulse response tail, e.g., k_{max} = 10 × T_{RMS}/T_s.

Path loss in a WLAN

Another key consideration is the issue of operating range relative to path loss. This issue plays an important role for determining the size of overlapping WLAN cells and distribution of APs. Path loss calculations are equally important for determining the radio's receiver sensitivity and transmitting power level and signal-to-noise ratio (SNR) requirements. As radios transmit signals to other receivers in a given area, the signal attenuates as a square of the distance D. The distance is the radius of a WLAN cell, as shown in Figure 16–4. The wavelength *lambda* is the ratio between the speed of light and the signal frequency. As the receiver moves away from the transmitter, the receiver's signal power decays until it reaches the receiver's noise floor, at which time the bit error rate (BER) becomes unacceptable. For indoor applications beyond 20 feet, propagation losses increase at about 30 dB per 100 feet. This loss occurs because of a combination of attenuation by walls,

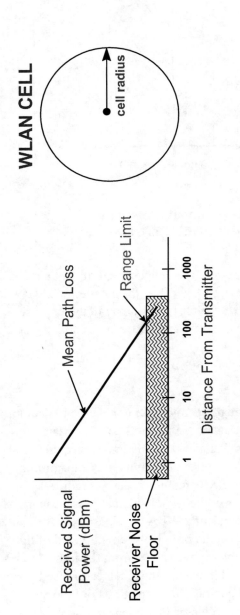

Figure 16–4: Free space path loss model

WLAN CELL

cell radius

Received Signal Power (dBm)

Mean Path Loss

Range Limit

Receiver Noise Floor

Distance From Transmitter

1   10   100   1000

Path Loss (dB) = 20 Log$_{10}$ (4 x PI x D/Lambda)

Where:

r = D is the radius of the WLAN cell

Lamda = c/f

where : c = speed of light (3 x 10^{8}ms^{-1})

f = signal frequency in Hz

ceilings, and furniture. Each wall constructed with sheet rock and wood typically attenuates the signal by 6 dB, and walls constructed with cement block attenuate the signal by 4 dB. However, additional losses may occur depending on the fading characteristics of the operating environment, which we describe in the next section. The same path loss principles apply for all frequency bands. However, as the operating frequency increases from 2.4 GHz to 5 GHz, for example, an additional path loss of 5 dB to 10 dB occurs. This situation results in a smaller cell radius and may require additional overlapping cells and APs to guarantee the same area as a system operating at 2.4 GHz.

Multipath fading

Another key consideration is the path loss due to multipath fading. Multipath fading occurs when the reflected signal paths refract off people, furniture, windows, and walls and scatter the transmitted signal. For example, moving the receiver from the transmitter a small distance, even only a few inches, can produce an additional loss of signal power on the order of 20 dB or more. Multipath fading is viewed as two separate factors and described as probability distribution functions. The first factor is a characteristic known as *log normal fading*, which is related to the coefficient products that result as the signal reflects off surfaces and propagates to the receiver. As the signal coefficient products propagate to the receiver, they are summed together with the direct path. Where they cancel each other, they cause significant attenuation of the transmitted signal. This effect is the second factor, known as *Rayleigh fading*. As previously mentioned, rake architectures and equalization are techniques used to correct for these effects.

Es/No vs BER PERFORMANCE

System performance tradeoffs are often made in the decision process when selecting a modulation type and data rate. System tradeoffs in terms of receiver sensitivity, range, and transmit power become every important for developing low-cost implementations, especially for higher rate 2.4 GHz IEEE 802.11b systems. Figure 16–5 illustrates a comparison of the theoretical Es/No vs BER curves for uncoded QPSK, PBCC 5.5–11 Mbit/s, CCK

5.5–11 Mbit/s, and Barker 1 and 2 Mbit/s. The theoretical curves include additive white Gaussian (AWG) noise in the channel. These curves are provided as a guide to assess the performance for a complete system implementing CCK and PBCC. However, to get better understanding of the overall systems performance, other factors such as multipath, signal fading, carrier phase noise, noise figure (NF), and other implementation losses should be considered in the link budget as part of the systems analysis.

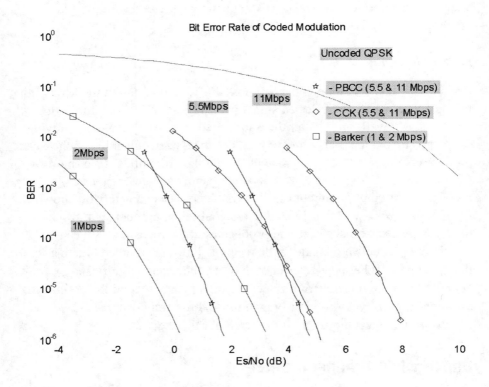

**Figure 16–5:   Theoretical Eb/No vs BER with AGW
for 2.4 GHz IEEE 802.11b**

DATA RATE VS AGGREGATE THROUGHPUT

The IEEE 802.11 standard defines data rate in terms of symbol rate or available bit rate. The PPDU data is modulated and transmitted over the RF or IR medium at this rate. This rate is often confused with the aggregate data throughput. The aggregate data rate takes into account the overhead associated with protocol frame structure, collisions, and implementation processing delays associated with frames processed by mobile STAs and APs. Simulations may be run in software to estimate the aggregate throughput of the protocol and benchmarked against IEEE 802.11 WLANs. However, calculating the aggregate throughput can be complex because there are a number of detailed variables to consider. The protocol overhead includes parameters such as RTS, CTS, and ACK frames; short, distributed, and priority interframe space (SIFS, DIFS, PIFS) timing; beacon periods and random backoff periods; estimated collisions; PPDU frame size; and RF propagation delays. A good rule of thumb for estimating the average aggregate throughput of an IEEE 802.11 wireless network is 75% of the data rate for distributed coordination function (DCF) operation and 85% of the data rate for point coordination function (PCF).

WLAN INSTALLATION AND SITE SURVEY

Many installations begin with a site survey. A site survey serves a number of purposes. First, the survey is used to determine the maximum operating range between an AP (i.e., fixed location) and mobile STAs for a specified transmit RF power level. Second, the survey helps identify holes of coverage due to multipath, interference sources, and neighboring existing WLAN installations. Third, it is used in cell planning of overlapping basic service areas (BSAs) and for layout of APs giving them hardwired access to existing wired Ethernet LAN infrastructures.

Today, many equipment manufactures have tests built into their products to conduct such surveys. PC laptops with IEEE 802.11 WLAN adaptor cards with embedded software tools are commonly used. In some cases, a spectrum analyzer with special directional antennas is used to measure path loss through walls and other obstructions and to pinpoint and identify interference sources. Some of the tests include BER, packet error rate (PER), and link quality measurements as a function of range. Typically, the tests are recorded using a pair of WLAN adaptors; one is set up in a fixed location and the other as a mobile STA. Every environment is different, and the number of APs required for a given installation depends upon the number of holes in the coverage area due to multipath and signal attenuation through walls, ceilings, and floors. However, on average, for indoor operation, the maximum operating distance between a mobile STA and an AP operating in the 2.4 GHz frequency and transmitting at an RF transmit power of +20 dBm (100 mW) at data rates of 1 Mbit/s and 2 Mbit/s yields approximately 400 ft and at the data rate of 11 Mbit/s, 100 ft.

INTERFERENCE IN THE 2.4 GHZ FREQUENCY BAND

The microwave oven used in household and commercial kitchens is the main interference source in the 2.4 GHz unlicensed frequency band. The magnetron tubes used in the microwave ovens radiate a continuous-wave (CW-like) interference that sweeps over tens of megahertz of the 2.4–2.483 GHz band during the positive half cycle of ac line voltage. The microwave oven's equivalent isotropically radiated power (EIRP) has a maximum ranging between 16 dBm and 33 dBm. The power cycle frequency is 50 Hz 20 ms or 60 Hz 16 ms depending upon the geographical location. In North America, the ac line frequency is 60 Hz, and the microwave oven's magnetron pulses on for 8 ms and off for 8 ms. The maximum packet length defined in the IEEE 802.11 protocol was designed to operate between the 8 ms pulses of the microwave energy.

Other sources of interference include neighboring in-band radios. Two types of interference are considered here. First is co-channel interference, which is induced from radios from adjacent cells that are on the same channel frequency. Proper cell planning of the channel frequency and hopping patterns and careful layout of the APs can minimize this interference. The second type of interference is from other systems such as neighboring DSSS and FHSS WLANs. Built into the IEEE 802.11 standard are three mechanisms used to help minimize the amount of interference. The first is the clear channel assessment (CCA), where the MAC protocol provides a method of collision avoidance. The second is processing gain, which provides some protection from FHSS radios, whose spectrum appears as narrowband interferers. The third is the hopping patterns; there is sufficient frequency spacing between pseudo-random hops to minimize the interference due to neighboring DSSS channels. To some degree, legacy 2.4 GHz IEEE 802.11 FHSS and DSSS systems and IEEE 802.11b high-rate WLANs do coexist. However, careful cell planning will help minimize the amount of interference a system will experience especially at the outer fringe of the cell.

ANTENNA DIVERSITY

Historically, antenna diversity has been an effective low-cost alternative solution used to combat and mitigate the effects of multipath and delay spread in WLAN radio receivers. It is relatively easy to implement in the mobile STAs and APs and does not require the signal processing hardware used in other diversity techniques. The object behind antenna diversity is to space the antennas apart from each other to minimize the effects of the uncorrelated multipath at the receiver. Spacing the antennas far apart allows the receiver to pick and demodulate the larger signal of the two signals. For IEEE 802.11 2.4 GHz implementations, the bit length of the preamble sync fields was selected based on these criteria. The antennas are typically spaced anywhere from 0.251 to several lambdas (i.e., wavelengths) apart. The amount of separation depends upon the amount of delay-spread tolerance required for a system to operate in a given operating environment. Adding antenna diversity will improve the PER performance of a wireless link by 2 to 1 as well as improve the availability of the link. There are a number of 2.4 GHz antennas on the market today with different configurations. Patch antennas are commonly used at the mobile STA PCMCIA implementations because of cost and size constraints. On the other hand, omni-directional antennas are used at the AP because they provide the optimal antenna coverage. Although antenna diversity is an option in the IEEE 802.11 standard, as a minimum, antenna diversity should always be considered at the AP as shown in Figure 16–6. This form of diversity will minimize the risk of packet loss due to multipath and interference and ensure optimal throughput performance in a system.

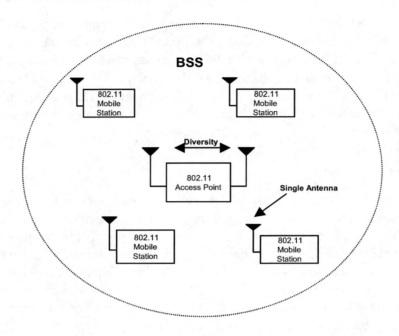

Figure 16–6:   Antenna diversity at the AP as a minimum

POWER MANAGEMENT DEFINED

Power management is key feature for all mobile STAs (i.e., clients) and APs.
It is often viewed as a key differentiator between radio chipset suppliers and
manufacturers of IEEE 802.11 Wi-Fi STA and AP products in the market
today. Power management has become equally important in APs. Now with
the availability of power over Ethernet, for enterprise deployments, AP
manufactures are designing products using the same design constraints as
mobile devices. Many of the mobile IEEE 802.11 Wi-Fi devices in the market
are dual-band, multimode, and feature rich. The demand for data rates
≥ 54 Mbit/s delivering maximum range and longer battery life are driving
chipset suppliers to consider innovative ways to design power amplifier
devices and consider new radio techniques, such as MIMO. We can expect
this trend to continue as the adoption of IEEE 802.11 Wi-Fi devices become
ubiquitous in the future. Power consumption and battery life are critical for

IEEE 802.11 Wi-Fi products used in platforms such as personal digital assistants (PDAs), Voice Over Internet Protocol (VoIP) phones, cell phones, wireless digital television (DTV), streaming video, and mobile video web pads. Historically, there has been confusion between the chipset suppliers and product manufacturers when specifying power consumption for the various modes of operation because they are not defined or specified in the IEEE 802.11 standard. Listed below are definitions of the modes of operation for IEEE 802.11 Wi-Fi mobile STAs (i.e., client devices). These definitions have become recognized by the industry as the modes of operation for which power consumption is specified for an IEEE 802.11 Wi-Fi radio. See Figure 16–7. The same holds true for APs and other application-specific devices.

a) **Transmit mode:** In this mode, only the transmit path in the radio is turned on. This includes digital circuits in the MAC, memory, host interface, the modulator functions in the baseband processor and the mixed-signal analog automatic gain control (AGC), active filters, and RF circuits, including the synthesizer PLL and power amplifier. During this mode of operation, data are transferred from the host to the MAC where the data are formatted and encapsulated into IEEE 802.11 packets; modulated and up-converted to either a 2.4 GHz RF carrier for IEEE 802.11b or IEEE 802.11g devices or a 5 GHz RF carrier for IEEE 802.11a, IEEE 802.11j, or IEEE 802.11h devices; and transmitted over the air. The amount of power consumed during transmit is highly dependent upon the conducted RF output power and the power-added efficiency of the power amplifiers. A power-added efficiency of 10% is common for most Class A and Class A/B power amplifiers and is the dominate contributor in power consumption during this mode of operation.

b) **Receive mode:** In this mode, only the receiver in the radio is turned on. This includes digital circuits in the MAC, host interface, memory, the demodulator functions in the baseband processor and the mixed-signal analog AGC, active filters, and RF circuits, including the synthesizer PLL. During this mode of operation, incoming data is received from the RF medium by the antenna at a carrier frequency of either 2.4 GHz for IEEE 802.11b and IEEE 802.11g devices or 5 GHz for IEEE 802.11a devices. The carrier frequency is down-converted to baseband

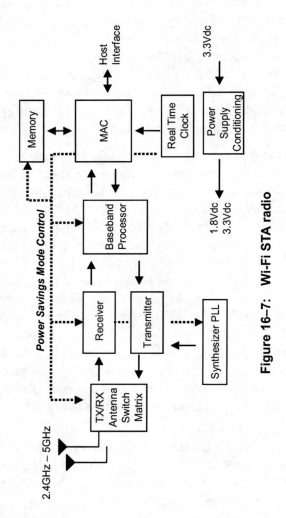

Figure 16–7: Wi-Fi STA radio

and digitally demodulated into serial data stream. The MAC captures valid IEEE 802.11 packets and passes them to the host.

c) **Idle mode:** In this mode, the receiver in the radio wakes from the sleep mode and turns on to listen and receive beacons. After the listening period for beacons expires, the radio transitions back to sleep. The power consumed is duty cycle depended and includes the power consumption in the receiver only when the receiver is turned on while listening for beacons. Many IEEE 802.11 radios have the capability to skip and not listen to every beacon, and this feature further reduces the power consumption while in this mode.

d) **Sleep mode:** In this mode, the entire radio is shut down, including the synthesizer and PLL, with the exception of the real-time clock. The real-time clock is programmed for a specific time to wake up the radio to listen for beacons or to transmit or receive data. During sleep, all of the information in memory is retained. The recovery time of the synthesizer is critical and needs to be carefully managed in order to minimize the possibility of violating the protocol timing for beacons and for receiving or transmitting data.

e) **Deep sleep mode:** In this mode, the entire radio is shut down; this mode is analogous to having your STA card that is plugged into a PC card or mini-PCI card slot disabled. The power being consumed during this time is the leakage draining from the ICs on the card and is usually specific in microwatts. The radio needs to be programmed when the radio is powered up. Such programming is usually performed by the combination of the host driver software and embedded MAC firmware.

The power consumption in the support circuitry in the radio should be considered when specifying the power consumptions for each mode of operation. The power-conditioning circuits for power supplies is sometimes missed in the budget. Many of the radio reference designs use low dropout regulators and switch power supplies to regulate and condition the power supplies (e.g., 1.8 V dc and 3.3 V dc), which are on the order of 85% to 90% efficient.

Glossary

Access point (AP): a special function provided by a particular type of IEEE 802.11 station (STA) that provides connectivity between mobile IEEE 802.11 STAs and a network infrastructure that may include both wired and wireless devices.

Ad hoc traffic indication message (ATIM) window: a period of time, immediately after the target beacon transmission time in an ad hoc WLAN, when the only permitted transmissions are ATIM frames that announce the availability of a frame buffered at one station (STA) in an ad hoc WLAN for another STA in that WLAN.

Ad hoc WLAN: a WLAN that has no access points (APs) and, typically, exists for a short period of time for a specific purpose, as at a meeting of a few people to share information.

Authentication: the act of proving the identity of a device or user of a device.

Authenticator: one protocol endpoint that executes IEEE 802.1X Extensible Authentication Protocol over LAN (EAPOL) with its peer endpoint, the Supplicant.

Beacon frame: a management frame that synchronizes the timers of the members of a basic service set (BSS), announces the identity of the extended service set, indicates required and supported configurations of the BSS, and communicates additional protocol information.

Ciphertext: the result of performing an encryption operation on some information.

Cleartext: information provided as input to an encryption operation.

Contention window (CW): the period of time after the conclusion of an IEEE 802.11 transmission, during which access attempts to the medium are randomized to reduce the probability of collisions.

Current contention window: the size of the interval (measured in slot times) from which a station (STA) chooses a random backoff value.

Delay spread: the difference in the time of arrival of a signal sent from a single source that travels over multiple paths to a receiver.

Distribution service: the function of an access point (AP) that determines how to deliver a frame to or from a station (STA) in its basic service set (BSS) to the frame's destination, typically including translation of a frame between LAN protocols and, perhaps, encapsulating the content of frame to preserve protocol information that would otherwise be lost in translation, e.g., Subnetwork Access Protocol (SNAP) encapsulation of the content of an Ethernet frame to preserve the value of the Ethertype field when sending to an IEEE 802.11 STA.

Distribution system (DS): the term in IEEE 802.11 for whatever connects one access point (AP) to another and provides the ability to deliver frames from a station (STA) in one basic service set (BSS) to a STA in a different BSS, typically an Ethernet LAN.

Enhanced distributed channel access (EDCA): an access mechanism built on the original, basic access mechanism that provides additional prioritization of frames.

Hidden station (STA): a STA that, because of its location relative to other STAs or because of obstacles to signal propagation, is not able to hear or be heard by all other STAs in a basic service set (BSS).

Hybrid coordinated channel access (HCCA): an access method that provides negotiated access agreements between two STAs.

Interframe space (IFS): a fixed period of time used by the IEEE 802.11 protocol to separate protocol operations. Currently, there are five IFS types defined: short interframe space (SIFS), priority interframe space (PIFS), arbitration interframe space (AIFS), distributed interframe space (DIFS), and extended interframe space (EIFS).

Multipath: a characteristic of electromagnetic signal propagation where a signal travels by many paths between a transmitter and receiver.

Nonce: a number that is used only once for a protocol negotiation, typically to demonstrate cryptographic liveness, i.e., that an earlier protocol exchange is not simply being repeated.

Robust security network (RSN): a WLAN using only the authentication and key management protocols (AKMPs) and cipher suites defined by IEEE 802.11i.

Slot time: the unit of measurement, defined uniquely for each physical layer (PHY), for backoff counter used during the contention window (CW).

Supplicant: one protocol endpoint that executes IEEE 802.1X Extensible Authentication Protocol over LAN (EAPOL) with its peer endpoint, the Authenticator.

Target beacon transmission time: a periodic time, well-known and easily calculated by every station (STA) in a basic service set (BSS), at which the Beacon frame will be transmitted if there are no other transmissions currently on the medium and the medium has been idle for the required interframe space to allow access.

Time unit (TU): 1024 μs.

Transition security network (TSN): a WLAN using both wired equivalent privacy (WEP) as well as the authentication and key management protocols (AKMPs) and cipher suites defined by IEEE 802.11i.

Transmit classification (TCLAS): a set of rules that is used to identify a specific set of frames as belonging to a transmit specification.

Transmit opportunity (TXOP): a point in time, optionally a periodic point in time, at which a station (STA) can begin a transmission or transmissions that can occupy the medium for a duration that has been negotiated with the access point (AP).

Transmit specification (TSPEC): a negotiated agreement between a station (STA) and an access point (AP) that identifies a traffic flow, specifies its characteristics, and determines when it will be allowed to use the medium.

Index

Numerics

2.4 GHz frequency band, 273, 327–331
  See also HR/DSSS
4-Way Handshake, 36, 42–43, 99, 106, 111–113, 130–131, 135, 219
5 GHz frequency band, 192, 273, 284, 327–331
  See also OFDM

A

aAirPropagationTime, 297–299
AC parameter record fields, 174
Access Policy subfield, 178
ACI subfield, 175
ACI/AIFSN subfield, 174
ACK control frame, 33, 42–43
ACK Policy subfield, 152–153
ACM subfield, 175
ACR, 309–310
Action Details subfield, 160, 194
Action management frame
  IEEE 802.11h, 192–196
  IEEE P802.11e extensions, 154, 159–169
  type and subtype values, 33
Action subfield, 160, 162, 167
Ad hoc WLAN, 349
ADDBA Request (Block ACK Action) frame, 167
ADDBA Response (Block ACK Action) frame, 167–168
Additive white Gaussian (AWG) noise, 340
Address fields, 37–39, 47
Address filtering, 227–228
ADDTS request (QoS Action) frame, 162
ADDTS response (QoS Action) frame, 162
Adjacent channel
  interference, 258, 281, 289, 304–306, 321, 328

nonoverlapping, 294, 308
operation on, 258, 289
overlapping, 259, 281, 290, 294, 328
rejection (ACR), 309–310
AES CCMP, 122–126
Aggregate data rate, 341
Aggregation subfield, 178
AID field, 56
AIFSN subfield, 175
AKM Suite Count field, 100
AKM Suite List field, 100
AKM Suites Table, 132–133
AKMP, 105–113, 219
aMPDUMaxLength, 247
Antenna diversity as WLAN design consideration, 343–344
APSD mechanism, 187
APSD subfield, 61, 77–78, 156, 178
Association Request management frame, 33, 54, 158
Association Response management frame, 33, 54, 158
Association service, 11, 14, 225–227
Association state, 12–14
ATIM management frame, 33, 55, 229–230
ATIM window, 74, 229–230, 349
ATIM Window field, 74
Attack countermeasures, 122
Australia, 271, 283, 304
Authentication
  algorithm, 65
  MAC management, 223–225
  original service, 10, 12–15
  reason codes, 62–64
  state, 12
  status codes, 65
Authentication Algorithm Number field, 56
Authentication management frame, 33, 36, 53, 65
Authentication Transaction Sequence Number field, 56
Automatic gain control (AGC), 274, 345

Short Slot Time subfield, 61
SIFS interval, 24, 27, 41–45, 146, 148,
 153, 175, 182–183, 311, 326
Signal field, 252, 285, 287, 300–302
Simple data frame, 45, 47–48
Site survey as WLAN design consideration,
 341–342
Sleep mode, 347
Slot time, 309, 311, 326, 351
Sniffer, 240
SNR, 71, 329, 337
Source address (SA), 39
Spain, wireless operation in, 207, 260,
 265–266, 272, 283–284, 294
Specification Interval field, 82, 181
Spectrum management, 200–205
Spectrum Management subfield, 60
Spurious emissions (Japan), 306–307
SSID information element, 70
STA
 management attributes, 241–246
 services, 10
Starting Sequence Control field, 146
State variables and services, relationship
 between (illustration), 13
Status Code field, 64–67, 240
Subband triplet (Country information
 element), 210–212, 296
Subtype subfield, 31
Supported Channels information element,
 63, 66, 199
Supported Rates information element,
 70–71
Surplus Bandwidth Allowance field, 180
Surplus Bandwidth field, 79
Suspension Interval field, 78, 179
SYNC field, 252, 255, 261, 267, 285,
 287–288, 319, 343
Synchronization, 232–235, 237

T

Tails bits, 275
Target beacon transmission time, 351
TBTT, 53, 233–234
TCLAS, 351

TCLAS information element, 80–82
TCLAS Processing information element,
 84–85
TCP/UDP IP classifier parameters, 81
Temporal key (TK), 106
TID subfield, 150
TIM information element, 72–74
Timer synchronization
 FH PHYs, 235
 IBSS, 234–235
 infrastructure BSS, 233–234
Timestamp field, 68
Timing intervals, 23–24
TKIP, 116–122
To DS subfield, 31, 35
TPC, 197–200, 284
TPC Report information element, 196,
 198–199
TPC Request information element, 196,
 198–199
Traffic identifier (TID), 146
Traffic Type subfield, 178
Transmit masks, 304–306, 321
Transmit mode, 345
Transmit power requirements
 DSSS, 258–261
 OFDM, 281–283
 operation in Japan, 302–304
Transmit sequence counter (TSC), 118
Transmitter address (TA), 39
Transmitter clock lock, 316
TS Delay information element, 84
TS Info field, 175
TSF, 201, 232–235
TSF timer, 237
TSID subfield, 178
TSInfo ACK Policy subfield, 178
TSINFO field, 77
TSN, 96, 100, 115–116
TSPEC, 145, 351
 See also EDCA
TSPEC information element, 77–79,
 175–180
Two-way handshake, 36, 106
TXOP, 142, 144–145, 351
TXOP Limit subfield, 175

Additional Books in the
IEEE Standards Wireless Networks Series

Wireless Communications Standards: A Study of IEEE 802.11™, 802.15™, and 802.16™ is the only IEEE book of its kind that covers all of the current 802 wireless standards!

Wireless Multimedia: A Handbook to the IEEE 802.15.3™ Standard clarifies the IEEE 802.15.3 standard for individuals who are implementing compliant devices and shows how the standard can be used to develop wireless multimedia applications.

Low-Rate Wireless Personal Area Networks: Enabling Wireless Sensors with IEEE 802.15.4™ is an excellent companion to the standard for those interested in the field of "simple" wireless connectivity with a further focus on wireless sensors and actuators for the industry in general.

Also coming in 2005....

* ***The IEEE Wireless Dictionary***
* ***WirelessMAN™ : Inside the IEEE 802.16™ Standard for Wireless Metropolitan Area Networks***

Visit: **http://standards.ieee.org/standardspress/**